MOLECULAR COLLISIONS IN THE INTERSTELLAR MEDIUM

Second Edition

In the interstellar medium – which occupies the space between the stars in galaxies – new stars are born from material that is replenished by the debris ejected by stars when they die. This book presents a detailed account of the atomic and molecular processes which give rise to the radiation we observe from the interstellar medium, knowledge that is essential to understanding star formation in our own and other galaxies.

This Second Edition has been thoroughly updated and extended to cover related topics in radiation theory. It considers the chemistry of the interstellar medium, both in the present epoch and the early Universe. The book discusses the physics and chemistry of shock waves, which are produced by the jets of matter generated as a consequence of star formation. The methods for calculating rates of collisional excitation of interstellar molecules and atoms are explained, with emphasis on the quantum mechanical method.

A comprehensive manual for studying collisional and radiative processes in the interstellar medium, this book will be ideal for researchers involved in calculating the rates of such processes, for those studying the interstellar medium and star formation, as well as for physical chemists specializing in collision theory or in the measurement of the rates of collision processes.

DAVID FLOWER is a professor of physics at the University of Durham, UK. He is a fellow of the Royal Astronomical Society and an editor of the Society's astronomy research journal, *Monthly Notices*. His research interests include atomic and molecular physics in astrophysical environments, and the physics of the interstellar medium.

Cambridge Astrophysics Series

Series editors

Andrew King, Douglas Lin, Stephen Maran, Jim Pringle and Martin Ward

Titles available in the series

7. Spectroscopy of Astrophysical Plasmas
   edited by A. Dalgarno and D. Layzer
10. Quasar Astronomy
    by Daniel W. Weedman
18. Plasma Loops in the Solar Corona
    by R. J. Bray, L. E. Cram, C. Durrant, R. E. Loughhead
19. Beams and Jets in Astrophysics
    edited by P. A. Hughes
22. Gamma-ray Astronomy 2nd Edition
    by P. V. Ramana Murthy, A. W. Wolfendale
24. Solar and Stellar Activity Cycles
    by Peter R. Wilson
26. X-ray Binaries
    by Walter H. G. Lewin, Jan van Paradijs, Edward P. J. van den Heuvel
27. RR Lyrae Stars
    by Horace A. Smith
28. Cataclysmic Variable Stars
    by Brian Warner
29. The Magellanic Clouds
    by Bengt E. Westerlund
32. Accretion Processes in Star Formation
    by Lee Hartmann
33. The Origin and Evolution of Planetary Nebulae
    by Sun Kwok
34. Solar and Stellar Magnetic Activity
    by Carolus J. Schrijver, Cornelis Zwaan
35. The Galaxies of the Local Group
    by Sidney van den Bergh
36. Stellar Rotation
    by Jean-Louis Tassoul
37. Extreme Ultraviolet Astronomy
    by Martin A. Barstow, Jay B. Holberg
38. Pulsar Astronomy 3rd Edition
    by Andrew G. Lyne, Francis Graham-Smith
39. Compact Stellar X-ray Sources
    edited by Walter H. G. Lewin, Michiel van der Klis
40. Evolutionary Processes in Binary and Multiple Stars
    by Peter Eggleton
41. The Physics of the Cosmic Microwave Background
    by Pavel D. Naselsky, Dmitry I. Novikov, Igor D. Novikov
42. Molecular Collisions in the Interstellar Medium 2nd Edition
    by David Flower

MOLECULAR COLLISIONS IN THE INTERSTELLAR MEDIUM
Second Edition

DAVID FLOWER
University of Durham

CAMBRIDGE
UNIVERSITY PRESS

CAMBRIDGE UNIVERSITY PRESS
Cambridge, New York, Melbourne, Madrid, Cape Town, Singapore, São Paulo

Cambridge University Press
The Edinburgh Building, Cambridge CB2 8RU, UK

Published in the United States of America by Cambridge University Press, New York

www.cambridge.org
Information on this title: www.cambridge.org/9780521844833

© D. Flower 2007

This publication is in copyright. Subject to statutory exception
and to the provisions of relevant collective licensing agreements,
no reproduction of any part may take place without
the written permission of Cambridge University Press.

First published 1990
First paperback edition 2003
Second edition 2007

Printed in the United Kingdom at the University Press, Cambridge

A catalogue record for this publication is available from the British Library

ISBN-13 978-0-521-84483-3 hardback

Cambridge University Press has no responsibility for the persistence or
accuracy of URLs for external or third-party internet websites referred to
in this publication, and does not guarantee that any content on such
websites is, or will remain, accurate or appropriate.

To Michael Seaton, mentor and friend

Contents

1	**Interstellar molecules**	**1**
1.1	Introduction	1
1.2	Chemistry in interstellar clouds	3
1.3	Chemical bistability in dense clouds	9
2	**Interstellar shocks and chemistry**	**12**
2.1	Introduction	12
2.2	The MHD conservation equations	13
2.3	The structure of interstellar shock waves	21
2.4	Shock waves in dark clouds	29
2.5	Shock waves in diffuse clouds	33
3	**The primordial gas**	**36**
3.1	Introduction	36
3.2	The governing equations	36
3.3	The role of molecules	39
3.4	Chemistry	42
3.5	Gravitational collapse	44
4	**The rotational excitation of molecules**	**49**
4.1	Introduction	49
4.2	The Born–Oppenheimer approximation	49
4.3	The scattering of an atom by a rigid rotator	52
4.4	The rotational excitation of non-linear molecules	69
5	**The vibrational excitation of linear molecules**	**82**
5.1	Introduction	82
5.2	The scattering of an atom by a vibrating rotor	82
5.3	Excitation of H_2 and HD in collisions with H_2 molecules	92
5.4	Cooling functions	93
6	**The excitation of fine structure transitions**	**98**
6.1	Introduction	98
6.2	Theory of fine structure excitation processes	99

7 Radiative transfer in molecular lines — 118
- 7.1 Introduction — 118
- 7.2 The radiative transfer equation — 119
- 7.3 The OH radical — 124
- 7.4 Producing population inversion — 128
- 7.5 Rotational excitation of OH by H_2 — 129

8 Charge transfer processes — 139
- 8.1 Introduction — 139
- 8.2 The Landau–Zener model — 140
- 8.3 The 'orbiting' model — 143
- 8.4 The quantum mechanical model — 145
- 8.5 Selective population of excited states — 151

9 Electron collisions — 153
- 9.1 Introduction — 153
- 9.2 Selection rules and LS-coupling — 154
- 9.3 Electron collisional excitation — 156
- 9.4 Resonances — 158
- 9.5 Forbidden line emission from Herbig–Haro objects — 161

10 Photon collisions — 163
- 10.1 Introduction — 163
- 10.2 The oscillator strength — 163
- 10.3 The transition probability — 165
- 10.4 Photoionization and radiative recombination — 166
- 10.5 Radiative transitions in molecules — 169

Appendix 1 *The atomic system of units* — 172
Appendix 2 *Reaction rate coefficients* — 173
References — 177
Index — 185

1
Interstellar molecules

1.1 Introduction

Our perception and understanding of the interstellar medium have been transformed over the last approximately 50 years. By the 1950s, optical and 21 cm radio measurements had demonstrated the presence of gas containing predominantly atomic hydrogen, together with a few atoms and ions (Na, K, Ca^+, Ti^+, Fe) that had absorption lines in the visible part of the spectrum. The gas was cold ($T \approx 100$ K) and had a number density $n(H) \approx 1$ cm^{-3}. 'Molecules' were represented by the radicals CH, CH^+ and CN, which have absorption bands in the visible. A dust component was known to be present, most directly from its obscuration of starlight. Over the last half-century, the construction of new telescopes, both ground-based and on board satellites, and rapid developments in receiver technology have led to over 100 molecular species being identified in the interstellar medium (see Table 1.1).

Foremost among the molecular species is the most abundant molecule in the Universe, H_2. Molecular hydrogen was first detected in absorption in the ultraviolet part of the spectrum, through its $X^1\Sigma_g^+ - B^1\Sigma_u^+$ Lyman electronic bands, by means of a sounding rocket [1]. It has since been observed at much higher spectral resolution by the Copernicus [2] and the FUSE [3] satellites. H_2 has also been observed in the far infrared by the *Infrared Space Observatory* (ISO) satellite [4], and through atmospheric windows in the near infrared. Thus, electronic, rotational and rovibrational transitions of H_2 have all been observed in the interstellar medium.

Cold H_2, at $T \lesssim 30$ K, maps the molecular clouds in galaxies but remains difficult to observe, being detectable only by means of ultraviolet electronic absorption lines in the Lyman and Werner ($X^1\Sigma_g^+ - C^1\Pi_u$) bands from above the Earth's atmosphere. Furthermore, a background continuum ultraviolet source, such as a hot star, is necessary for such observations. As scattering and absorption by dust are pronounced at short wavelengths, such observations are limited by the extinction along the line of sight through the molecular cloud and towards the star. Consequently, CO, which is the next most abundant interstellar molecule after H_2 [$n(CO)/n(H_2) \approx 10^{-4}$], has been used extensively as a tracer of molecular material. CO is readily observable from the ground at millimetre wavelengths; its longest wavelength rotational transition, $J = 1 \rightarrow 0$, falls at 2.6 mm. Indeed, radiative transitions between the low-lying rotational levels of CO are major contributors to the cooling of molecular clouds.

In this initial chapter, my aim is to summarize key aspects of our current understanding of the chemistry of the interstellar medium. We shall consider 'dense' (or 'dark') molecular clouds, whose interiors are shielded from the Galactic ultraviolet radiation field by dust absorption in the outer layers. This background radiation field owes its existence to hot (e.g. O and B) stars that emit in the ultraviolet. Although those photons beyond the H I

Table 1.1. *Observed interstellar molecular species, ordered according to complexity, including isotopes and isomers (August 2004). Courtesy NRAO website: http://www.cv.nrao.edu/~awootten/allmols.html*

Diatomic	Triatomic	4-atoms	5-atoms	6-atoms	7-atoms
H_2	C_3	c-C_3H	C_5	C_5H	C_6H
AlF	C_2H	l-C_3H	C_4H	l-H_2C_4	CH_2CHCN
AlCl	C_2O	C_3N	C_4Si	C_2H_4	CH_3C_2H
C_2	C_2S	C_3O	l-C_3H_2	CH_3CN	HC_5N
CH	CH_2	C_3S	c-C_3H_2	CH_3NC	$HCOCH_3$
CH^+	HCN	C_2H_2	CH_2CN	CH_3OH	NH_2CH_3
CN	HCO	CH_2D^+	CH_4	CH_3SH	c-C_2H_4O
CO	HCO^+	HCCN	HC_3N	HC_3NH^+	CH_2CHOH
CO^+	HCS^+	$HCNH^+$	HC_2NC	HC_2CHO	
CP	HOC^+	HNCO	HCOOH	NH_2CHO	
CSi	H_2O	HNCS	H_2CHN	C_5N	
HCl	H_2S	$HOCO^+$	H_2C_2O		
KCl	HNC	H_2CO	H_2NCN		
NH	HNO	H_2CN	HNC_3		
NO	MgCN	H_2CS	SiH_4		
NS	MgNC	H_3O^+	H_2COH^+		
NaCl	N_2H^+	NH_3			
OH	N_2O	SiC_3			
PN	NaCN				
SO	OCS				
SO^+	SO_2				
SiN	c-SiC_2				
SiO	CO_2				
SiS	NH_2				
CS	H_3^+				
HF	SiCN				
SH	AlNC				
FeO					

8-atoms	9-atoms	10-atoms	11-atoms	13-atoms
CH_3C_3N	CH_3C_4H	CH_3C_5N	HC_9N	$HC_{11}N$
$HCOOCH_3$	CH_3CH_2CN	$(CH_3)_2CO$		
CH_3COOH	$(CH_3)_2O$	NH_2CH_2COOH		
C_7H	CH_3CH_2OH	CH_3CH_2CHO		
H_2C_6	HC_7N			
CH_2OHCHO	C_8H			
CH_2CHCHO				

Lyman limit at 91.2 nm are absorbed, by atomic hydrogen, in the immediate vicinities of the hot stars, photons of longer wavelengths propagate into the general interstellar medium, where they are ultimately absorbed by either the dust or the gas. Accordingly, we shall also consider 'diffuse' (or 'translucent') clouds, which are traversed by this radiation field. The discussion will concentrate on reactions in the gas phase, although it is clear that grain-surface reactions are significant in the production of some species, notably H_2. Indeed, reactions between H atoms on the surfaces of grains are the only effective means of forming molecular hydrogen in the interstellar medium.

The only significant sources of ionization in dark clouds are cosmic rays and, possibly, X-rays in the vicinity of sources of such radiation. Cosmic rays with energies of a few MeV ionize hydrogen, producing 'secondary' electrons with energies of typically 30 eV [5]. The secondary electrons can collisionally excite the Lyman and Werner electronic transitions of H_2, leading to either the dissociation of the molecule or, more probably, to radiative cascade back to the electronic ground state; these processes are considered in Chapter 10. The fluorescence photons have wavelengths in the ultraviolet part of the spectrum and are sufficiently energetic to ionize and dissociate a number of species in the gas. The significance of this radiation field, generated *within* 'dark' clouds, was first recognized by Prasad and Tarafadar [6], and the photon spectrum was subseqently evaluated in detail [7].

1.2 Chemistry in interstellar clouds

1.2.1 Formation of molecular hydrogen

Even in the so-called 'dense' interstellar clouds, particle number densities are extremely low compared with those at atmospheric pressure, at which $n \approx 10^{19}$ cm^{-3}, and conditions are far from those in thermodynamic equilibrium at the kinetic temperature of the gas. Hydrogen exists predominantly in its 1s ground state, and collisions between hydrogen atoms can proceed along either of two potential energy curves, in which the electronic spins are either parallel (triplet state) or anti-parallel (singlet state). As the individual electronic orbital angular momenta are zero, the resultant orbital angular momentum and its projection on the internuclear axis of the quasi-molecule are also zero; the corresponding molecular states are denoted $^3\Sigma$ and $^1\Sigma$. As is well known from the Heitler–London theory of the H_2 molecule, the $^1\Sigma$ state is attractive whereas the $^3\Sigma$ state is repulsive. Figure 1.1 illustrates the variation of these potential energy curves with internuclear distance, R.

As the atoms are initially unbound, their total energy, E, is positive (the zero of the total energy is taken at the molecular dissociation limit). In order to stabilize, the system must lose energy and E become negative. In the gaseous phase, this may occur:

- by means of three-body collisions, the third body taking away the excess energy, or
- by means of radiative processes.

Three-body collisions are extremely improbable at interstellar densities, and so the only way in which the system can stabilize is through the emission of a photon. However, transitions between the $^3\Sigma$ and $^1\Sigma$ electronic potential energy curves are forbidden to electric dipole radiation as they involve a change in the total spin quantum number. Radiative transitions involving the nuclear degrees of freedom (rotation and vibration) are also forbidden, as the H_2 molecule is homonuclear and does not possess a permanent dipole moment. (Radiative selection rules will be considered in Chapter 10). It follows that the formation of H_2 by two-body association in the gas phase cannot explain the observed presence of molecular hydrogen in the interstellar medium.

An alternative and still the only viable theory of H_2 formation in interstellar clouds is through grain-surface reactions [8]. The grain acts as a catalyst, playing the role of the third body in a three-body reaction. Early estimates of the rate of formation of H_2 on grains [9] suggested that most of the hydrogen would be expected to be in molecular form in dense molecular clouds.

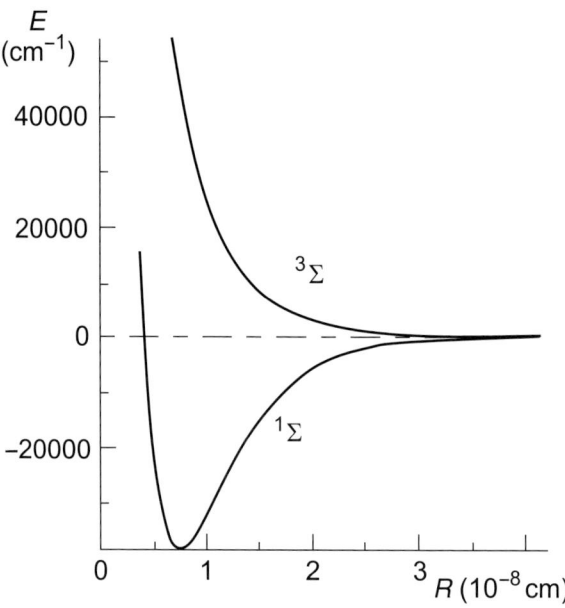

Figure 1.1 Electronic potential energy curves correlating with H(1s) + H(1s) at large internuclear separation R. The notation is $^{2S+1}\Lambda$, where $S = 0$ or 1 is the total electronic spin and Λ is the projection of electronic orbital angular momentum on the internuclear axis. States with $\Lambda = 0$, as here, are denoted Σ.

In the more recent literature (e.g. [10]), a distinction has been made between the Eley–Rideal, or 'prompt', and the Langmuir–Hinshelwood, or 'diffusion', modes of formation of H_2 on grains. In the former case, the collision of an H atom from the gas with an H atom which is already physi-bound to the grain surface leads to the formation of H_2 and its release into the gas. In the latter case, the second H atom migrates across the grain surface by thermal hopping until it encounters and reacts with another H atom, releasing the molecule. It remains unclear whether one or the other of these mechanisms dominates. Also uncertain is the partition of the binding enery of the H_2 (4.48 eV) following the reaction, between translational, vibrational and rotational modes of the molecule and excitation of the phonon spectrum of the grain. Both experimental and theoretical work is continuing in an attempt to elucidate these issues. A recent theoretical study of the Eley–Rideal mechanism [11], in which the polycyclic aromatic hydrocarbon (PAH) coronene was assumed to be the substrate, indicates that most of the energy released in the reaction goes into vibrational excitation of the product molecule. These issues are important not only for the thermal balance of the medium – the kinetic energy of the H_2 is a significant heat source – but also for predictions of the spectrum of H_2 emitted from regions in which it forms, such as photon-dominated regions (PDRs) and behind dissociative shock waves propagating in molecular gas.

1.2.2 Formation of molecules more complex than H_2

Ion-neutral reactions play key roles in the formation of molecules in the interstellar medium. In the context of dense molecular clouds, such reactions were first discussed in detail by Herbst and Klemperer [12]. The long-range attraction, due to the polarization

1.2 Chemistry in interstellar clouds

of the molecule by the ion, ensures that these reactions are generally rapid at the low temperatures of the ambient medium, providing they are exothermic. In shocked molecular gas, the latter restriction no longer applies, owing to the higher kinetic temperature and, in magnetohydrodynamic shocks, to ion-neutral streaming associated with ambipolar diffusion (see Chapter 2).

Dense clouds

In dense molecular clouds, cosmic ray (cr) ionization of hydrogen is the primary ionization process,

$$H_2 + cr \rightarrow H_2^+ + e^- + cr \tag{1.1}$$

followed rapidly by

$$H_2^+ + H_2 \rightarrow H_3^+ + H \tag{1.2}$$

which yields H_3^+. The H_3^+ ion has a propensity to transfer a proton to neutral species with higher proton affinities than H_2. Heavier molecules may then be produced through a sequence of ion-neutral and dissociative recombination reactions:

$$H_3^+ + O \rightarrow OH^+ + H_2 \tag{1.3}$$
$$OH^+ + H_2 \rightarrow H_2O^+ + H \tag{1.4}$$

followed by

$$H_2O^+ + e^- \rightarrow OH + H \tag{1.5}$$

or

$$H_2O^+ + H_2 \rightarrow H_3O^+ + H \tag{1.6}$$

and

$$H_3O^+ + e^- \rightarrow H_2O + H \tag{1.7}$$

The hydroxyl radical and the water molecule are produced in this sequence of reactions. A similar sequence, initiated by

$$H_3^+ + C \rightarrow CH^+ + H_2 \tag{1.8}$$

generates the carbon-bearing species CH, CH_2, CH_3 and CH_4. Dissociative recombinations such as (1.5) and (1.7) are generally rapid, owing to the long-range coulomb attraction between the reactants involved and the low mass of the electron, which ensures a high collision frequency.

Oxidation of the products of the above hydrogenation reactions results in the formation of CO, e.g.

$$CH + O \rightarrow CO + H \tag{1.9}$$

and O_2 can form in the reaction

$$OH + O \rightarrow O_2 + H \tag{1.10}$$

Both OH and O_2 react with C to produce CO. In oxygen-rich environments ($n_O > n_C$), most of the carbon is converted into CO when molecules form.

Observations with the *Submillimeter Wave Astronomy Satellite* (SWAS) satellite [13, 14] have shown that H_2O and O_2 are much less abundant in dense clouds than expected on the basis of gas-phase chemical models. At least some of these molecules are expected to be frozen on to the grains at the low temperatures that prevail in such clouds.

Unlike the corresponding reactions involving C and O, the protonation of N by H_3^+ is endothermic and has a negligible rate at low kinetic temperatures, T. Molecular nitrogen is believed to be produced in the reaction

$$N + OH \rightarrow NO + H \tag{1.11}$$

followed by

$$NO + N \rightarrow N_2 + O \tag{1.12}$$

Proton transfer from H_3^+ to N_2 then yields the molecular ion N_2H^+. N_2 can also react with He^+, which is produced by cosmic-ray ionization of He:

$$He^+ + N_2 \rightarrow N^+ + N + He \tag{1.13}$$

This may be followed by the slightly endothermic reaction

$$N^+ + H_2 \rightarrow NH^+ + H \tag{1.14}$$

which initiates a sequence of hydrogenation reactions with H_2, terminating at NH_4^+. At low T, reaction (1.14) is driven principally by ortho-H_2, whose ground state lies approximately 170 K above that of para-H_2 and increases the enthalpy of the reactants by the corresponding amount.[1] The dissociative recombination of NH_4^+

$$NH_4^+ + e^- \rightarrow NH_3 + H \tag{1.15}$$

produces ammonia, NH_3.

Ammonia can be produced also on grains and be subsequently released into the gas phase. Reactions on grain surfaces are believed to occur predominantly with adsorbed H, which, being light, can migrate relatively rapidly across the grain surface. The saturated (in hydrogen) species CH_4, NH_3 and H_2O can form in this way.

Atomic sulphur reacts with H_3^+, producing SH^+. However, the reactions of SH^+ and H_2S^+ with H_2 are endothermic and proceed with negligible rates in cold gas. SH^+ may undergo (slow) radiative association with H_2

$$SH^+ + H_2 \rightarrow H_3S^+ + h\nu \tag{1.16}$$

---

[1] The energies of the reactants and the products of a reaction are expressed as *enthalpies* and often given in units of kcal mol^{-1} (1 kcal mol$^{-1} \equiv 503.4$ K); units of kJ mol$^{-1} \equiv 120.3$ K are also used.

1.2 Chemistry in interstellar clouds

and H_2S and SH are produced in the subsequent dissociative recombination of H_3S^+. On the other hand, SH^+ may also undergo charge transfer with S

$$SH^+ + S \rightarrow SH + S^+ \tag{1.17}$$

and SH reacts with O

$$SH + O \rightarrow SO + H \tag{1.18}$$

and SO reacts with C

$$SO + C \rightarrow CS + O \tag{1.19}$$

CS is recognized as a tracer of dense, cold gas. Unlike CO, the compound of carbon with the corresponding element (O) in the the second row of the Periodic Table, CS has a large permanent dipole moment. As a consequence, the populations of the rotational energy levels of CS begin to thermalize at densities that are two orders of magnitude higher than is the case for CO. The populations of the excited states increase quadratically with the gas density below the critical density, at which the rates of radiative and collisional de-excitation become equal, but only linearly above the critical density. Thus, the emission from CS increases, relative to the emission from CO, at gas densities higher than the critical density for CO, but lower than the critical density for CS. The chemical routes to simple interstellar molecules, outlined above, are shown in Fig. 1.2.

In dense clouds, H_3^+ reacts mainly with CO:

$$H_3^+ + CO \rightarrow HCO^+ + H_2 \tag{1.20}$$

However, there is indirect observational evidence that, in protostellar cores, where the temperature may be very low ($T \lesssim 10$ K), essentially all molecules containing species heavier than He may have frozen on to the grains. Under these circumstances, HD (which, like H_2, forms on the surfaces of grains) may participate in the sequence of reactions

$$H_3^+ + HD \rightarrow H_2D^+ + H_2 \tag{1.21}$$

$$H_2D^+ + HD \rightarrow D_2H^+ + H_2 \tag{1.22}$$

and

$$D_2H^+ + HD \rightarrow D_3^+ + H_2 \tag{1.23}$$

which can result in high degrees of *fractionation* of deuterium, i.e. enhanced fractional abundances of the the deuterated isotopes of H_3^+, relative to the elemental abundance ratio n_D/n_H. Fractionation occurs because the zero-point vibrational energies of successive ions in the reaction sequence, H_3^+, H_2D^+, D_2H^+ and D_3^+, decrease owing to the increase in their masses. At such low temperatures, the extent of the fractionation might be such that D_3^+ is the most abundant ion in the gas.

8   *Interstellar molecules*

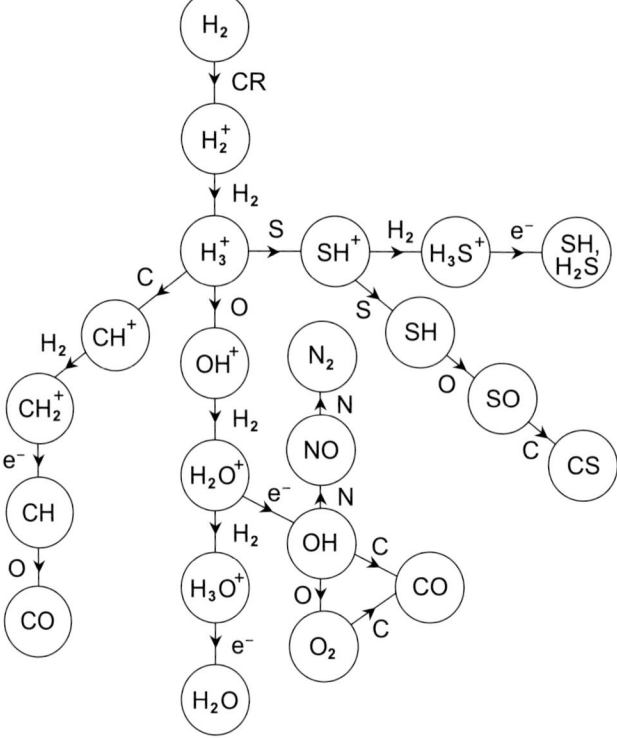

Figure 1.2 Chemical routes to simple interstellar molecules in dense molecular clouds.

Diffuse clouds

The diffuse interstellar gas is permeated by photons with energies $h\nu < 13.6$ eV, the ionization potential of atomic hydrogen. Consequently, atoms with lower ionization potentials, such as C, Si, S and Fe, are photoionized. This process not only modifies the reactants available to form molecules in the gas phase but also produces a higher fractional abundance of electrons. Furthermore, some of the molecules and radicals that are formed can be dissociated, or even ionized, by the background radiation field. The energy input from the radiation field leads to higher kinetic temperatures ($T \approx 80$ K) than those prevailing in dark clouds.

The ionization potential of atomic oxygen is higher than that of atomic hydrogen, but by only a small amount, equivalent to 227 K. The charge transfer of O with H^+, which is produced by cosmic ray ionization of H,

$$H^+ + O \rightarrow H + O^+ - 227 \text{ K} \tag{1.24}$$

is consequently endothermic by 227 K but nonetheless proceeds quite rapidly in diffuse clouds. The hydrogen abstraction reaction with H_2

$$O^+ + H_2 \rightarrow OH^+ + H \tag{1.25}$$

1.3 Chemical bistability in dense clouds

then initiates the same sequence which occurs in dense clouds [reactions (1.4)–(1.7) above]. The corresponding reaction of C^+ with H_2

$$C^+ + H_2 \rightarrow CH^+ + H - 4640 \text{ K} \tag{1.26}$$

has a large endothermicity (4640 K). It has been proposed that CH^+ forms in shocked gas, as a consequence of interstellar turbulence, or as a result of diffusion. None of these proposals has proved completely successful in accounting for both the observed column densities and the line profiles and shifts of CH^+ and its associated radical, CH.

1.3 Chemical bistability in dense clouds

The chemical composition of molecular gas attains steady state if the chemical timescales are shorter than the dynamical and other relevant timescales that characterize the medium. The assumption of steady state provides a point of reference, but the abundances of some of the chemical species in dark clouds are unlikely to reach a time-independent equilibrium. Under these circumstances, their current abundances depend on their 'initial' values.

Even if steady state has been attained, it does not guarantee that the abundances are independent of the initial conditions. The first suggestion that abundances in steady state may depend on the initial conditions is to be found in the work of Graedel *et al.* [15], who discovered the existence of two distinct chemical phases. This discovery remained neglected for a decade, partly as a consequence of the belief at that time in a very low value for the coefficient of dissociative recombination of H_3^+ with electrons – a belief which subsequent experimental work [16, 17] showed to be unfounded. As we have already seen, H_3^+ plays a pivotal role in interstellar chemistry, and its rate of recombination with electrons is crucially important. The two chemical phases were rediscovered in the early 1990s [18], and subsequent work [19] demonstrated the relevance of the phenomenon of bistability to the chemistry of dark clouds.

'Bistability', or chemical hysteresis, is analogous to the phenomenon of magnetic hysteresis in ferromagnetic materials. The degree of ionization in the steady state of gas with densities representative of dark clouds is plotted in Fig. 1.3; each point in this figure is a steady-state solution. The solutions display an unstable branch, between the points A and B, and two stable branches, which are labelled 'HIP' and 'LIP' for the high and the low ionization phases, respectively. In the range of gas density $2 \times 10^3 \leq n_H \leq 2 \times 10^4$ cm^{-3}, bistability occurs. The degree of ionization is approximately an order of magnitude greater in the HIP than in the LIP.

The difference in the degree of ionization in the HIP and LIP is reflected in the abundances of many atomic and molecular species. For example, the corresponding values of the ratio $n(C)/n(CO)$ are plotted in Fig. 1.4; this ratio is of the order of 10^{-1} in the HIP but less than 10^{-2} in the LIP.

An alternative and instructive way of presenting the results is illustrated in Fig. 1.5, which shows $n(O_2)$ plotted agains $n(C)$ for models in which $n_H = 2 \times 10^4$ cm^{-3}. In steady state, the LIP solution yields a relatively high molecular oxygen and low atomic carbon abundance, and vice versa for the HIP solution. Slightly different initial conditions can lead to the the opposite (LIP or HIP) solution prevailing in steady state. Thus, small variations in parameters

10   *Interstellar molecules*

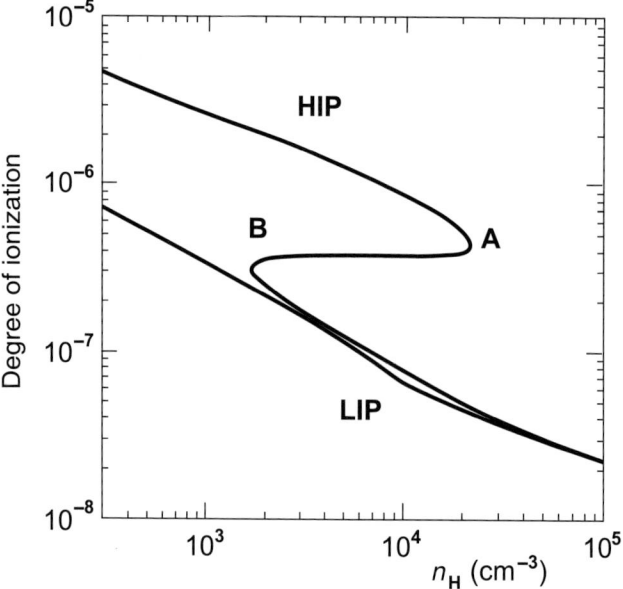

Figure 1.3 Degree of ionization of interstellar gas, computed in steady state, for a range of values of the gas density $n_H \approx n(H) + 2n(H_2)$. The high and low ionization phases are indicated. The curve has an unstable branch between points A and B, where it displays the phenomenon of bistability.

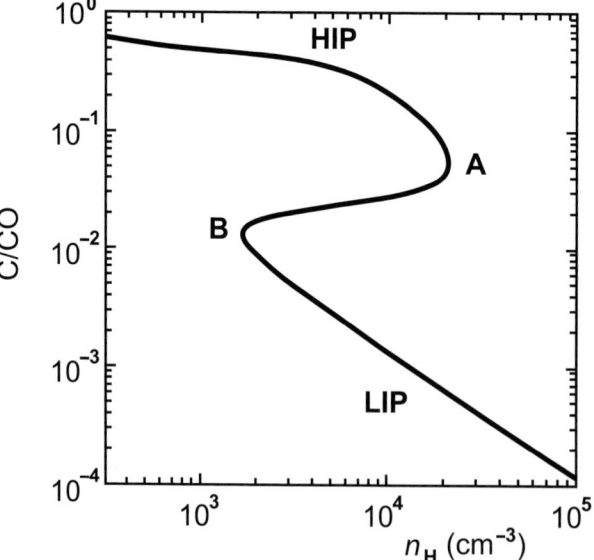

Figure 1.4 Bistability in the variation of $n(C)/n(CO)$ with n_H.

1.3 Chemical bistability in dense clouds

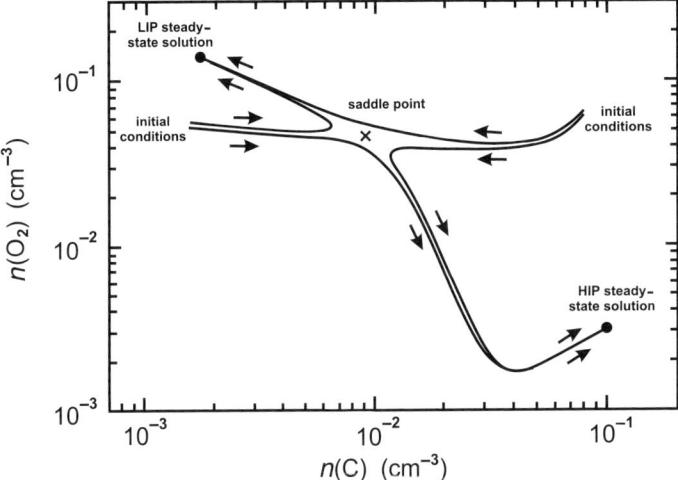

Figure 1.5 Evolution to steady state from various initial conditions. Small changes in the initial conditions can lead to very different steady-state solutions.

such as the local gas density or the cosmic ray flux might be responsible for the gas evolving chemically to one, rather than the other, of these two distinct chemical phases.

As we have seen, the reaction (1.3) initiates the formation of oxygen-bearing species in dense clouds. Hence, a high equilibrium abundance of H_3^+ is associated with a low O abundance and high abundances of species such as O_2 and H_2O. The latter are scavengers of atomic ions, particularly H^+ and C^+, forming molecular ions that can be neutralized rapidly by dissociative recombination with electrons; these conditions are those of the LIP. If the fractional electron density increases for some reason, such as a decrease in the gas density or an increase in the cosmic ray ionization rate, a point is reached at which the rate of removal of H_3^+ by recombination with electrons becomes equal to the rate of its removal in protonation reactions. With the lower H_3^+ abundance are associated a higher O abundance and lower abundances of O_2 and H_2O. The consequent reduction on the rates of neutralization of H^+ and C^+ enhances the electron density further, and a transition occurs in the other (HIP) phase. Thus, molecular clouds display chemical instabilities that lead to transitions between chemical phases. This phenomenon may be compared to the occurrence of thermal instabilities in the interstellar medium, which lead to the existence of thermal phases [20], and is equally important.

In fact, the interstellar gas is being continually perturbed by dynamical phenomena that give rise to shock waves and also to turbulence. Interstellar clouds that are static and have constant density and temperature do not exist, although this fact does not necessarily imply that models based on these assumptions are irrelevant – rather, that the models are more or less good approximations to reality. In Chapter 2, we consider the effects of shock waves propagating within interstellar clouds and, in particular, their influence on the chemistry.

2

Interstellar shocks and chemistry

2.1 Introduction

Considerable effort has been directed at studying the formation of molecular species in the ambient interstellar medium. By 'ambient' is to be understood the cold gas whose equilibrium temperature is determined by the balance of heating through the action of cosmic or X-rays and of the interstellar background radiation field at wavelengths $\lambda > 91.2$ nm. Models have been made of both diffuse and dense clouds, using both chemical equilibrium and time-dependent codes. There have been many publications in this area; they all owe much to the pioneering studies of Herbst and Klemperer [12] and Black and Dalgarno [21].

In the ambient medium, the kinetic temperature of the gas is low ($T \lesssim 100$ K) and even mildly endothermic processes, such as the important charge exchange reaction

$$H^+ + O \to H + O^+ - 227\,K \tag{2.1}$$

are effectively inhibited. The gas-phase chemistry of the ambient medium is dominated by reactions with no endothermicity and no reaction barrier.

Observations of molecular clouds have established beyond reasonable doubt that shocks propagate within them. Hyperthermal molecular line widths and emission from highly excited states of molecules, particularly H_2 and CO, indicate that shock wave heating is occurring in the Orion molecular cloud, for example. Shocks, driven by turbulent motions in molecular clouds and the expansion of compact H II regions, or by the jets of matter associated with the formation of proto-stars, are undoubtedly a common phenomenon.

While passage through a shock wave is a mechanism for producing vibrationally excited H_2 molecules [22–24], calculations show that, for shock speeds $u_s \gtrsim 20$ km s^{-1}, the H_2 would be dissociated by thermal collisions on passing through the shock front. As line width measurements indicated shock speeds much in excess of 20 km s^{-1} (e.g. [25]), there appeared to be an inconsistency in the shock-excitation hypothesis.

In early work on hydromagnetic shock waves in the interstellar medium [26], the ionized and neutral fluids were assumed to be fully coupled. In this case, the action of the magnetic field (on the ionized gas) is simultaneously transmitted to the neutral gas. Field et al. [26] recognized that, if the degree of ionization is low, appreciable decoupling of the flows of the ionized and the neutral fluids might ensue. The magnetic field may be considered to be 'frozen' in the ionized fluid, which is electrically conducting. When the magnetic field has a component perpendicular to the flow direction, the compression of the ionized gas is accompanied by the compression of the magnetic field; but the effects of this compression are felt by the neutrals, via collisions with the ions, only after a delay. Differences develop in

2.2 The MHD conservation equations

the flow speeds of the charged and neutral fluids, a phenomenon which is termed 'ion-neutral drift', from the viewpoint of the fluids, or 'ambipolar diffusion', from the viewpoint of the magnetic field (which diffuses through the neutrals, together with the ions).

Effects associated with the partial decoupling of the charged and neutral fluids were subsequently investigated by Mullan [27]. As a consequence of ambipolar diffusion, the neutral fluid is compressed and heated in advance of the shock discontinuity. This heating mechanism was incorporated, in an approximate manner, into the shock model of Hollenbach and McKee [28], but it was not until the work of Draine [29] that a quantitative model of such magnetohydrodynamical (MHD) shock waves became available. The model of Draine, like its predecessors, assumed that steady state had been attained; only recently have time-dependent models – which describe the temporal evolution as well as the spatial structure of shock waves – found their way into the literature [30–33].

As a result of shock wave heating, chemical reactions, which are endothermic or which present reaction barriers attaining a few tenths of an eV, become significant. The requisite energy derives from the motion of the shock wave (i.e. from the 'piston' that drives it) and is transmitted to the gas in the form of thermal energy and, in MHD shocks, through the ion-neutral drift. Thus, reactions such as

$$C^+ + H_2 \rightarrow CH^+ + H - 4640 \text{ K} \tag{2.2}$$

and

$$O + H_2 \rightarrow OH + H \tag{2.3}$$

which has a barrier of 2980 K, are unimportant in the ambient medium but assume significance in regions which have undergone shock heating.

Just as the shock structure is important in determining the chemistry of the medium, so the chemistry is important to the structure of the shock wave. Chemical reactions affect directly the degree of ionization of the medium and hence the interaction with the magnetic field. Furthermore, chemical reactions influence the abundances of the atomic and molecular species which cool the gas, principally through the collisional excitation of rovibrational (Chapters 4 and 5) and fine structure (Chapter 6) transitions. The dynamical and chemical conservation equations are interdependent and must be solved in parallel. So too must the equations for the population densities of the rovibrational levels of the H_2 molecule, which is the main coolant of shocked molecular gas. There is a delay in the response of the level populations to changes in the density and the kinetic temperature of the gas; this delay is taken into account only when the equations for the level populations are integrated *in parallel* with the MHD and chemical conservation equations, to which we now turn.

2.2 The MHD conservation equations

The quantities with which we shall be concerned are the numbers of particles, their mass, momentum and energy. It has already been mentioned that the ionized and neutral fluids can develop different flow velocities; their temperatures may also differ. Furthermore, the temperatures of the ions and the electrons at any given point in the flow may not be equal. The development of differences in the velocities of the positively and negatively charged fluids is resisted by large electrical forces; these ensure that the velocities and the number densities of the positive and negative particles are effectively equal everywhere.

2.2.1 The equations in one dimension

In the analysis that follows, it will be assumed that a stationary state has been attained, in which case $\partial/\partial t = 0$. The total time derivative may then be written as

$$\frac{d}{dt} = \frac{\partial}{\partial t} + \mathbf{u} \cdot \nabla = \mathbf{u} \cdot \nabla \qquad (2.4)$$

where $\mathbf{u}$ is the flow velocity and $\nabla = \hat{\mathbf{i}}\partial/\partial x + \hat{\mathbf{j}}\partial/\partial y + \hat{\mathbf{k}}\partial/\partial z$ is the gradient operator. If the flow is plane-parallel in the z-direction, then $\nabla = \hat{\mathbf{k}}\partial/\partial z$ and

$$\frac{d}{dt} = u\frac{d}{dz} \qquad (2.5)$$

By making these assumptions, we exclude the possibility of studying rigorously both the temporal evolution of the shock wave and its structure in more than one spatial dimension; but we simplify considerably the numerical aspects of the problem, which reduces to solving coupled *ordinary* (rather than partial) differential equations.

Subject to the assumptions above, the equation for the number density of neutral particles states that

$$\frac{d}{dz}(\rho_n u_n/\mu_n) = \mathcal{N}_n \qquad (2.6)$$

where ρ_n is the mass density of the neutrals at the point z, μ_n their mean molecular weight, and u_n is their flow speed in the z-direction. The 'source' term on the right-hand side of equation (2.6) is the rate of creation (or destruction, if negative) of neutral particles per unit volume through recombination and ionization processes. For example, the formation of molecular from atomic hydrogen results in a net reduction in the number of neutral particles ($\mathcal{N}_n < 0$).

An analogous equation holds for the positively (and the negatively) charged fluid,

$$\frac{d}{dz}(\rho_+ u_+/\mu_+) = \mathcal{N}_+ \qquad (2.7)$$

The positively charged fluid, denoted by '+', comprises the positive ions and positively charged grains. In general, $\mathcal{N}_n \neq -\mathcal{N}_+$: dissociative recombination processes, for example,

$$CH^+ + e^- \rightarrow C + H \qquad (2.8)$$

result in the destruction of *one* ion but create *two* neutrals.

The equation of mass conservation for the neutrals may be written

$$\frac{d}{dz}(\rho_n u_n) = \mathcal{S}_n \qquad (2.9)$$

where $\mathcal{S}_n$ denotes the rate per unit volume at which neutral mass is created or destroyed. The corresponding equation for the positively charged fluid is

$$\frac{d}{dz}(\rho_+ u_+) = \mathcal{S}_+ \qquad (2.10)$$

2.2 The MHD conservation equations

As neutral mass may be created only through the destruction of charged mass, by recombination of positively and negatively charged particles, $\mathcal{S}_+ + \mathcal{S}_- = -\mathcal{S}_n$, where the subscript '$-$' refers to the negatively charged fluid, which comprises the electrons, negative ions and the negatively charged grains. When the negatively charged particle in the recombination reaction is an electron, which is most frequently the case, $\mathcal{S}_-$ is negligible.

Momentum must also be conserved; for the neutral fluid, the equation of momentum conservation takes the form

$$\frac{d}{dz}\left(\rho_n u_n^2 + \frac{\rho_n k_B T_n}{\mu_n}\right) = \mathcal{A}_n \tag{2.11}$$

where T_n is the temperature of the neutral fluid at the point z, and $\mathcal{A}_n$ determines the rate at which momentum is being gained or lost by unit volume of the neutral gas; k_B is Boltzmann's constant. Momentum transfer occurs principally in collisions of the neutrals with ions and with charged grains. As the positively and the negatively charged particles have the same number density and the same flow speed, their combined equation of momentum conservation may be written

$$\frac{d}{dz}\left[(\rho_+ + \rho_-)u_+^2 + \frac{\rho_+ k_B(T_+ + T_-)}{\mu_+} + \frac{B^2}{8\pi}\right] = -\mathcal{A}_n \tag{2.12}$$

where we have used the fact that $\rho_+/\mu_+ = \rho_-/\mu_-$, owing to the overall charge neutrality. In equation (2.12), B is the component of the magnetic field perpendicular to the z-direction, and $B^2/(8\pi)$ is the magnetic pressure term. The magnetic field acts directly on the charged fluid, which communicates its effects to the neutral fluid through collision processes. If the magnetic field is assumed to be 'frozen' into the charged fluid, the condition

$$Bu_+ = B_0 u_s \tag{2.13}$$

is satisfied; B_0 is the value of the magnetic field strength upstream of the shock wave, in the 'preshock' gas, and u_s is the shock speed. In the reference frame of the shock wave, in which the conservation equations are formulated, the initial ('upstream' or 'preshock') flow speeds of the neutral and charged fluids are both equal to the shock speed. Taking the flow to be in the positive z-direction implies that the shock wave is propagating in the negative z-direction.

The condition of energy conservation, applied to the neutral fluid, yields

$$\frac{d}{dz}\left[\frac{\rho_n u_n^3}{2} + \frac{5\rho_n u_n k_B T_n}{2\mu_n} + \frac{\rho_n u_n U_n}{\mu_n}\right] = \mathcal{B}_n \tag{2.14}$$

where U_n denotes the mean internal energy per neutral particle, and $\mathcal{B}_n$ is the rate of gain (or loss, if negative) of energy per unit volume of the neutral fluid. The internal energy consists essentially of the rovibrational excitation energy of the H_2 molecule, whose excited state population densities can become appreciable as the temperature and the density increase, owing to the passage of the shock wave.

In order to discriminate between the temperatures of the positively and negatively charged fluids, it is necessary to separate their energy conservation relations. Draine [34] derived these equations for the case of a MHD shock wave. However, the difference between

T_+ and T_- is difficult to evaluate accurately, and it has only a minor influence on the dynamical and chemical structure of the shock wave, in general. Accordingly, we consider the combined equation of energy conservation of the (positively and negatively) charged fluid, which takes the form

$$\frac{d}{dz}\left[\frac{(\rho_+ + \rho_-)u_+^3}{2} + \frac{5\rho_+ u_+ k_B(T_+ + T_-)}{2\mu_+} + \frac{u_+ B^2}{4\pi}\right]$$
$$= \mathcal{B}_+ + \mathcal{B}_- \qquad (2.15)$$

where $(\mathcal{B}_+ + \mathcal{B}_-)$ is the rate of energy gain per unit volume of the charged fluid.

2.2.2 The role of the magnetic field

We have already seen that the selective action of the magnetic field on the charged particles can give rise to different flow speeds for the charged and the neutral fluids. Consider first the neutral fluid. If a sonic point occurs in the flow, at the point where

$$u_n^2 = \frac{5k_B T_n}{3\mu_n} \equiv c_s^2 \qquad (2.16)$$

and c_s is the adiabatic sound speed, then the neutral flow becomes discontinuous and a shock occurs. In fact, this shock 'discontinuity' has a finite thickness, owing to viscous forces, which is of the same order as the length scale that characterizes elastic collisions between the neutral particles.

It is possible to integrate the conservation equations through the shock 'discontinuity' by introducing *artificial viscosity* terms [35]. Providing the transition from the pre- to the postshock gas occurs adiabatically, i.e. processes of energy transfer (notably radiative losses) are negligible, integration of the conservation equations through the 'discontinuity' automatically satisfies the Rankine–Hugoniot relations. The latter specify the compression ratio and the temperature ratio across the shock front, in the limit in which the shock front may be treated as a discontinuity; the Rankine–Hugoniot relations are derived in Section 2.3.1 below.

Similarly, a discontinuity occurs in the flow of the charged fluid at the point at which

$$u_+^2 = \frac{5k_B(T_+ + T_-)}{3(\mu_+ + \mu_-)} + \frac{B^2}{4\pi(\rho_+ + \rho_-)} \equiv c_m^2 \qquad (2.17)$$

where c_m is the magnetosonic speed in the charged fluid. As $u_+ \leq u_s$ everywhere, in the reference frame of the shock wave, a discontinuity in the flow of the charged fluid cannot occur if

$$u_s < c_m$$

a condition which is satisfied in magnetically dominated flows. It follows that flows can exist, which are discontinuous in only the neutral flow variables.

In practice, modest values of the magnetic field strength are sufficient to suppress the discontinuity in the flow variables of the charged fluid when the degree of ionization of the medium is low. For example, if $u_s = 10$ km s^{-1} and the charged mass density $\rho_+ + \rho_- = 2 \times 10^{-25}$ g cm^{-3} (corresponding to C$^+$ ions in a medium with a total particle density of about 50 cm^{-3}), $B \approx 1\,\mu$G is all that is required.

The region upstream of the discontinuity, in which the charged fluid has been compressed along with the magnetic field, has been termed the 'magnetic precursor' or 'acceleration zone'; the width of the precursor increases with the magnetic field strength. As the shock evolves, the discontinuity in the neutral flow moves progressively downstream and weakens, until finally the discontinuity is suppressed. Thus, for sufficiently large field strengths, the shock wave evolves from 'jump' or J-type, to J-type with a magnetic precursor, to 'continuous' or C-type [31]. To describe this evolution rigorously, a time-dependent MHD code must be used; but the evolution can be simulated by means of calculations, which are not explicitly time-dependent, as will be seen below. The mass density of the charged fluid, $\rho_+ + \rho_-$, is a factor determining the magnetosonic speed and hence the width of the acceleration zone. The endothermic reaction (1.26) of C$^+$ ions with H$_2$ molecules can be initiated by ion-neutral drift in this zone, yielding CH$^+$. The rapid exothermic reactions

$$CH^+ + H_2 \rightarrow CH_2^+ + H \tag{2.18}$$

and

$$CH_2^+ + H_2 \rightarrow CH_3^+ + H \tag{2.19}$$

then lead to the formation of CH$_2^+$ and CH$_3^+$. The molecular ions are destroyed by dissociative recombination reactions, such as

$$CH^+ + e^- \rightarrow C + H \tag{2.20}$$

which are believed to be rapid at the relevant temperatures. The net result is the neutralization of an important part of the ionized component of the gas. This process, of partial neutralization, can occur on a distance scale which is comparable with the dimensions of the acceleration zone, resulting in a significant enhancement of the width of this zone through the increase in the magnetic field term in equation (2.17).

In diffuse clouds, the situation is rendered more complicated by photoionization processes, notably

$$C + h\nu \rightarrow C^+ + e^- \tag{2.21}$$

which restore ions to the medium and which also occur over a characteristic distance scale that is comparable with the width of the acceleration zone. However, before considering further the chemistry in shock waves, in both diffuse and dense clouds, it is appropriate to consider in more detail the 'source terms' $(\mathcal{N}, \mathcal{S}, \mathcal{A}, \mathcal{B})$ appearing in the MHD conservation equations.

2.2.3 The source terms

The 'source terms' that appear on the right-hand sides of the MHD conservation equations presented above contain the micro-physics of the problem. These terms describe the interactions between the particles of the medium, including the grains, and the ways in which these interactions modify the number and mass densities, momentum and energy of the charged and neutral fluids. The principal terms will be introduced below; their hierarchy of importance depends on the context of the problem being considered. For a detailed discussion of the source terms, the reader may consult the paper by Draine [34].

Let us denote the net rate at which a particular atomic or molecular species, α, is produced per unit volume by $\mathcal{C}_\alpha$; a net destruction rate corresponds to $\mathcal{C}_\alpha < 0$. The total number of neutral particles produced per unit volume and time is

$$\mathcal{N}_n = \sum_{\alpha_n} \mathcal{C}_{\alpha_n} \qquad (2.22)$$

where the subscript 'n' identifies the species as being neutral. Similarly, for the positive ions

$$\mathcal{N}_+ = \sum_{\alpha_+} \mathcal{C}_{\alpha_+} \qquad (2.23)$$

As already noted, $\mathcal{N}_+ \neq -\mathcal{N}_n$, in general.

Creation of neutral mass proceeds at a rate per unit volume

$$\mathcal{S}_n = \sum_{\alpha_n} \mathcal{C}_{\alpha_n} m_{\alpha_n} \qquad (2.24)$$

where m_{α_n} is the mass of the neutral species α_n; the corresponding expression for the positive ions is

$$\mathcal{S}_+ = \sum_{\alpha_+} \mathcal{C}_{\alpha_+} m_{\alpha_+} \qquad (2.25)$$

In this case, $\mathcal{S}_+ + \mathcal{S}_- = -\mathcal{S}_n$ must be satisfied.

Momentum is transferred between the charged and the neutral fluids through ion-neutral reactions and elastic scattering. Denoting a particular ion-neutral reaction by β, we may write

$$\mathcal{C}_\alpha = \sum_\beta \mathcal{C}_{\alpha\beta} \qquad (2.26)$$

and the associated rate of momentum transfer from the charged to the neutral fluid is

$$\mathcal{A}_n^{(i)} = \sum_{\alpha_n \beta} \left[\sum_{\mathcal{C}_{\alpha_n\beta} > 0} \mathcal{C}_{\alpha_n\beta} m_{\alpha_n} u_\beta(\text{CM}) + \sum_{\mathcal{C}_{\alpha_n\beta} < 0} \mathcal{C}_{\alpha_n\beta} m_{\alpha_n} u_n \right] \qquad (2.27)$$

where the centre-of-mass collision velocity is given by

$$u_\beta(\text{CM}) = \frac{m_i u_\pm + m_n u_n}{m_i + m_n} \qquad (2.28)$$

where m_i is the mass of a (positive or negative) ion and m_n is the mass of the neutral. Equation (2.27) expresses the fact that, when a neutral is a product and hence $\mathcal{C}_{\alpha_n\beta} > 0$, it is created at the centre-of-mass velocity of the reaction β. On the other hand, when a neutral is a reactant ($\mathcal{C}_{\alpha_n\beta} < 0$), it is removed while moving with the velocity of the neutral fluid, u_n.

2.2 The MHD conservation equations

Osterbrock [36] derived an expression for the cross-section for momentum transfer in a collision between a charged and a neutral particle from considerations of the long-range charge-induced dipole interaction between the colliding pair; the expression that Osterbrock derived is

$$\sigma_{\text{in}} = 2.41\pi \left[\frac{e^2 \alpha_n}{m_{\text{in}} v_{\text{in}}^2} \right]^{\frac{1}{2}} \tag{2.29}$$

where e is the electron charge, α_n is the polarizability of the neutral, $m_{\text{in}} = m_i m_n/(m_i + m_n)$ is the reduced mass of the ion-neutral pair, and v_{in} is the relative collision speed. The expression (2.29) exceeds by 20% the 'Langevin' cross-section (cf. Chapter 8). Equation (2.29) has been shown to be valid at low collision speeds but to underestimate the momentum transfer at high collision speeds [37]. The polarizabilities of the principal constituents of the neutral fluid are: atomic hydrogen, $\alpha_H = 4.5\ a_0^3$; molecular hydrogen, $\alpha_{H_2} = 5.2\ a_0^3$; and helium, $\alpha_{He} = 1.4\ a_0^3$. The polarizabilities of H and H_2 are similar in magnitude and substantially larger than that of He.

The rate of transfer of momentum to the neutral fluid, owing to elastic scattering on the ions, is given by

$$\mathcal{A}_n^{(\text{ii})} = \frac{\rho_n \rho_i}{\mu_n + \mu_i} \langle \sigma v \rangle_{\text{in}} (u_i - u_n) \tag{2.30}$$

where

$$\langle \sigma v \rangle_{\text{in}} = 2.41\pi \left(\frac{e^2 \alpha_n}{\mu_{\text{in}}} \right)^{\frac{1}{2}} \tag{2.31}$$

is the corresponding rate coefficient; $\mu_{\text{in}} = \mu_i \mu_n/(\mu_i + \mu_n)$ is the reduced mass, evaluated using the mean molecular weights of the ionized and neutral fluids.

Momentum transfer between the neutral gas and charged grains is important in dense clouds, where the degree of ionization of the gas is low. In this case, the cross-section may be taken to be approximately equal to the the geometrical cross-section of the grain, πa_g^2, where a_g is the grain radius. (There is a correction to the geometrical cross-section, significant at low collision speeds, arising from the polarization of the neutral by the charged grain [38]). As the the collision speed is, to a good approximation, equal to the ion-neutral drift speed, $|u_i - u_n|$, and $\mu_g \gg \mu_n$, the rate of momentum transfer between the neutral gas and the charged grains (in dense clouds, most of the grains have a single negative charge) is

$$\mathcal{A}_n^{(\text{iii})} = \rho_n n_g \pi a_g^2 |u_i - u_n| (u_i - u_n) \tag{2.32}$$

The total rate of momentum transfer is $\mathcal{A}_n = \mathcal{A}_n^{(\text{i})} + \mathcal{A}_n^{(\text{ii})} + \mathcal{A}_n^{(\text{iii})}$.

Various physical processes lead to energy exchange between the charged and the neutral fluids. Chemical reactions are responsible for kinetic energy transfer from the charged to the neutral fluid, at a rate per unit volume

$$\mathcal{B}_n^{(\text{i})} = \sum_{\alpha_n \beta} \left[\sum_{C_{\alpha_n \beta} > 0} C_{\alpha_n \beta} \frac{1}{2} m_{\alpha_n} u_\beta^2 (\text{CM}) + \sum_{C_{\alpha_n \beta} < 0} C_{\alpha_n \beta} \frac{1}{2} m_{\alpha_n} u_n^2 \right] \tag{2.33}$$

An analogous expression holds for transfer of kinetic energy from the neutral to the charged fluid; the kinetic energy of the electrons may be neglected, in comparison with that of the ions.

When considering heat transfer between fluids, a distinction has to be made once again between formation ($C_{\alpha_n\beta} > 0$) and destruction ($C_{\alpha_n\beta} < 0$) processes. When an ion and an electron at temperatures T_+ and T_-, respectively, dissociatively recombine to form two neutrals, as in the reaction

$$\text{CH}^+ + \text{e}^- \rightarrow \text{C} + \text{H} \tag{2.34}$$

an amount of heat $3k_B(T_+ + T_-)/2$ is transferred to the neutral fluid. On the other hand, an amount of heat $3k_B T_n/2$ is lost by the neutral fluid through photoionization, as in

$$\text{C} + h\nu \rightarrow \text{C}^+ + \text{e}^- \tag{2.35}$$

The net rate of thermal energy transfer to the neutral fluid is

$$\mathcal{B}_n^{(ii)} = \sum_{\alpha_n\beta} \left[\sum_{C_{\alpha_n\beta}>0} C_{\alpha_n\beta} \frac{3}{2} k_B \frac{T_+ + T_-}{2} + \sum_{C_{\alpha_n\beta}<0} C_{\alpha_n\beta} \frac{3}{2} k_B \frac{T_n}{2} \right] \tag{2.36}$$

Dissociative recombination and photoionization are the most rapid and important processes determining the degree of ionization in shocks propagating in diffuse interstellar clouds. In the interiors of dense clouds, the degree of ionization is lower, owing to the absence of ionizing photons, which are absorbed and scattered by dust in the outer layers; cosmic ray ionization takes over but is a slow process.

When photoionization does occur, the corresponding heating rate is given by

$$\mathcal{B}_-^{(iii)} = \sum_\alpha n_\alpha \int_{\nu_\alpha}^{\nu_H} \frac{4\pi J_\nu}{h\nu} a_\nu(\alpha)(h\nu - h\nu_\alpha)\, d\nu \tag{2.37}$$

where J_ν is the mean radiation intensity at frequency ν, $a_\nu(\alpha)$ is the frequency-dependent photoionization cross-section of species α, and ν_α is the photoionization threshold frequency. The integral extends to the Lyman limit in atomic hydrogen, $h\nu = 13.598$ eV; photons of higher energy tend to be absorbed (by atomic hydrogen) in the immediate vicinities of the sources of the ultraviolet radiation.

Chemical reactions also affect the thermal balance of the medium by virtue of their energy defects, ΔE. The corresponding rate of heating of the neutral fluid is

$$\mathcal{B}_n^{(iv)} = \sum_{\alpha_n\beta} \sum_{C_{\alpha_n\beta}>0} C_{\alpha_n\beta} \frac{M_\beta - m_{\alpha_n}}{M_\beta} \Delta E_\beta \tag{2.38}$$

where M_β is the total mass of the products of reaction β, including m_{α_n}. The factor $(M_\beta - m_{\alpha_n})/M_\beta$ determines the partition of energy among the reaction products, with the lighter products carrying off more of the energy, ΔE_β, released in the reaction.

Elastic scattering of the neutrals on the ions results in the exchange of energy between the fluids. The neutral fluid is heated through this process at a rate given by

$$\mathcal{B}_n^{(v)} = \frac{\rho_n \rho_i}{\mu_n \mu_i} \langle \sigma v \rangle_{in} \frac{2\mu_n \mu_i}{(\mu_n + \mu_i)^2} [\frac{3}{2} k_B(T_i - T_n) + \frac{1}{2}(u_i - u_n)(\mu_i u_i + \mu_n u_n)] \tag{2.39}$$

where the rate coefficient for ion-neutral elastic scattering, $\langle\sigma v\rangle_{\rm in}$, is given by equation (2.31) above. Inspection of equation (2.39) shows that $\mathcal{B}_{\rm i}^{\rm (v)} = -\mathcal{B}_{\rm n}^{\rm (v)}$. The corresponding rate of energy transfer from the charged grains to the neutral fluid is

$$\mathcal{B}_{\rm n}^{\rm (vi)} = \rho_{\rm n} n_{\rm g} \pi a_{\rm g}^2 |u_{\rm i} - u_{\rm n}|(u_{\rm i} - u_{\rm n})u_{\rm i} \qquad (2.40)$$

where we assume $\mu_{\rm g} \gg \mu_{\rm n}$.

Collisional excitation, followed by radiative decay at an optically thin wavelength, is an important source of energy loss from the gas and must be taken into account. Particularly significant is the collisional excitation of rovibrational transitions in molecules and of fine structure transitions in atoms and ions, as discussed in Chapters 4–6. The rates of cooling processes are proportional to the number densities of the coolants, which depend in turn on the chemical reactions occurring within the shocked gas.

In shocks that give rise to appreciable collisional dissociation of molecular hydrogen the reformation of H$_2$ in the cooling flow of the shock wave represents a significant heating process. Molecular hydrogen forms on grains, and the H$_2$ molecules are returned to the gas phase with a finite amount of translational energy; this is subsequently converted to heat, through elastic collisions with the other constituents of the gas. The rate of heating is proportional to the rate of formation of H$_2$ and to the translational energy of the molecule as it leaves the grain. The total energy released when a molecule of hydrogen forms is 4.48 eV, the molecular binding energy. It is often assumed that this energy is partitioned, in equal fractions of 1/3, as internal (rovibrational) and translational energies of the molecule, and with 1/3 being recovered by the grain in phonon excitation. Whether this assumption is correct remains to be established, probably by means of experiments in the laboratory.

2.3   The structure of interstellar shock waves

In the previous section, we introduced the MHD conservation equations applicable to one-dimensional, steady-state, multi-fluid flows; these equations enable the structure of C-type shock waves to be calculated. However, shock waves in the interstellar medium are not necessarily, perhaps not normally, of C-type. When a shock wave is produced, for example in a collision between interstellar clouds at supersonic relative speed, the shock wave is initially of J-type. Depending on the shock speed and the magnetic field strength, this J-type shock wave may develop a magnetic precursor and ultimately become C-type.

The shock speed, $u_{\rm s}$, is an important parameter. The kinetic energy flux associated with the shock wave, $\rho u_{\rm s}^3/2$, increases rapidly with $u_{\rm s}$. Some of this energy is used to heat the gas, at the shock discontinuity. When the temperature, T, exceeds a few thousand degrees, molecular hydrogen begins to be collisionally dissociated. Because H$_2$ is a major coolant, its destruction leads to a further increase in T. Ultimately, the adiabatic sound speed, $c_{\rm s} = (\gamma k_{\rm B} T/\mu)^{\frac{1}{2}}$, where γ is the ratio of specific heats at constant pressure and volume, approaches the flow speed and a discontinuity occurs in the flow.

From the conservation relations presented in the previous section, the equations which are applicable to the 'discontinuity' in a J-type shock wave, and to the cooling flow behind the discontinuity, may be derived. As we have already noted, the so-called 'discontinuity' has a finite width, owing to the effects of viscosity and thermal conduction, which are characterized by length scales comparable with the mean free path for elastic scattering. The process of elastic scattering tends to equalize the values of parameters, such as u and T, associated with the flows of the various fluids. Accordingly, we shall assume single-fluid flow in what follows; but we note that this assumption is not valid for the grains, particularly the more massive grains, which have large inertia.

The conservation equations for single-fluid flow are obtained by adding the equations derived in the previous section for multi-fluid flow, i.e. for the neutral, positively and negatively charged fluids. The sums of the source terms appearing on the right-hand sides of the resulting equations of conservation of mass and momentum are identically zero. However, the number density and the energy of the flow are not conserved, in general. The number density can vary because of reactions, such as the collisional dissociation of H_2

$$H + H_2 \rightarrow H + H + H \tag{2.41}$$

in which there are two reactants but three products. Energy is lost from the flow in the form of radiation, as already mentioned. Thus, the conservation equations may be written in the form

$$\frac{d}{dz}\left(\frac{\rho u}{\mu}\right) = \mathcal{N} \tag{2.42}$$

$$\frac{d}{dz}(\rho u) = 0 \tag{2.43}$$

$$\frac{d}{dz}\left[\rho u^2 + \frac{\rho k_B T}{\mu} + \frac{B^2}{8\pi}\right] = 0 \tag{2.44}$$

and

$$\frac{d}{dz}\left[\frac{\rho u^3}{2} + \frac{5\rho u k_B T}{2\mu} + \frac{\rho u U}{\mu} + \frac{u B^2}{4\pi}\right] = \mathcal{B} \tag{2.45}$$

where

$$Bu = B_0 u_s$$

and where the subscript '0' denotes quantities in the preshock gas. Equivalently, we have [using equation (2.43)] that

$$\frac{d}{dz}\left(\frac{B}{\rho}\right) = 0 \tag{2.46}$$

when the magnetic field is frozen in the fluid.

The solution of equations (2.43), (2.44), (2.45) and (2.46) for the flow variables u, ρ, T and B, across the discontinuity and in the cooling flow, was considered by Field et al. [26]. If the molecular weight of the gas also varies, owing to reactions such as (2.41) above, equation (2.42) must also be included. We consider first the discontinuity, then the cooling flow.

2.3 The structure of interstellar shock waves

2.3.1 The 'discontinuity' in a J-type shock wave

We have already noted that the width of the 'discontinuity' is determined by viscosity and thermal conduction, and hence by the distance scale for elastic scattering processes. Chemical reactions, including collisional dissociation (2.41), and collisional processes leading to the emission of radiation, and hence cooling, are all *in*elastic processes, for which the characteristic distance scales are larger, typically by at least an order of magnitude, than the corresponding elastic scattering processes. Thus, within the shock 'discontinuity', the source terms $\mathcal{N}$ and $\mathcal{B}$ in equations (2.42) and (2.45), respectively, can be taken equal to zero. Furthermore, the flux of internal energy is constant, as the populations of internal energy states do not change. In the context of the equation of energy conservation, the shock transition ('discontinuity') may be qualified as adiabatic, i.e. there is no exchange of energy with the shock's environment. Viscosity and thermal conduction are significant only within the shock transition, where the velocity and thermal gradients are large; these processes (viscosity and thermal conduction) are not included in the above equations, as they can be neglected on either side of the 'discontinuity'. Thus, relations can be obtained between the flow variables immediately downstream and upstream of the discontinuity. These equations – commonly referred to as the *Rankine–Hugoniot relations* – are

$$\rho_1 u_1 = \rho_0 u_s \tag{2.47}$$

$$\rho_1 u_1^2 + \frac{\rho_1 k_B T_1}{\mu} + \frac{B_1^2}{8\pi} = \rho_0 u_s^2 + \frac{\rho_0 k_B T_0}{\mu} + \frac{B_0^2}{8\pi} \tag{2.48}$$

$$\frac{\rho_1 u_1^3}{2} + \frac{5\rho_1 u_1 k_B T_1}{2\mu} + \frac{u_1 B_1^2}{4\pi} = \frac{\rho_0 u_s^3}{2} + \frac{5\rho_0 u_s k_B T_0}{2\mu} + \frac{u_s B_0^2}{4\pi} \tag{2.49}$$

$$\frac{B_1}{\rho_1} = \frac{B_0}{\rho_0} \tag{2.50}$$

where the subscript '0' denotes the preshock gas, upstream of the discontinuity, '1' denotes the postshock gas, downstream of the discontinuity, and the molecular weight μ is constant. Equations (2.47) – (2.50) may be combined to yield a quadratic equation for the compression ratio, ρ_1/ρ_0, across the discontinuity:

$$2(2-\gamma)b\left(\frac{\rho_1}{\rho_0}\right)^2 + [(\gamma-1)M^2 + 2\gamma(1+b)]\frac{\rho_1}{\rho_0} - (\gamma+1)M^2 = 0 \tag{2.51}$$

In equation (2.51), M is the shock Mach number, the ratio of the shock speed, u_s, to the isothermal sound speed in the preshock gas, $(k_B T_0/\mu)^{\frac{1}{2}}$; $b = B_0^2/(8\pi p_0)$ is the ratio of the magnetic pressure, $B_0^2/(8\pi)$, to the gas pressure, $p_0 = \rho_0 k_B T_0/\mu$, in the preshock gas. The ratio of specific heats at constant pressure and volume, γ, should be taken equal to 5/3, the value appropriate to a gas with only translational degrees of freedom; this is because the internal energy of molecules such as H_2 does not have time to adjust to the changes in the temperature and density across the shock discontinuity. This adjustment occurs in the cooling flow, where the equations for the populations of the rovibrational levels of H_2 should be solved in parallel with the hydrodynamic conservation equations, in order to follow correctly the variation in the internal energy, U.

In the absence of a magnetic field ($B_0 = 0 = B_1$), the compression ratio across the shock discontinuity is given by

$$\frac{\rho_1}{\rho_0} = \frac{p_0 + h^2 p_1}{p_1 + h^2 p_0} \quad (2.52)$$

where $h^2 = (\gamma + 1)/(\gamma - 1)$. The pressure ratio across the shock discontinuity, p_1/p_0, is related to the isothermal Mach number in the preshock gas by

$$M^2 = \frac{\gamma + 1}{2} \frac{p_1}{p_0} + \frac{\gamma - 1}{2} \quad (2.53)$$

The processes that determine the thickness of the shock transition – viscosity and thermal conduction – are irreversible and give rise to an increase in entropy across the shock front. It can be shown that, as a consequence, the condition $p_1 > p_0$ must apply; it follows from (2.52) that $\rho_1/\rho_0 > 1$. Thus, both the pressure and the density of the gas *increase* as the gas traverses the discontinuity. The quadratic equation (2.51) has two roots, in general, but only one of the solutions corresponds to a compression shock, in which $\rho_1/\rho_0 > 1$.

Equation (2.52) shows that, in the limit of $p_1 \gg p_0$, $\rho_1/\rho_0 \to h^2 = 4$ for $\gamma = 5/3$. From equation (2.47), we see that, in this limit,

$$\frac{u_1}{u_s} = \frac{\rho_0}{\rho_1} = \frac{1}{4}$$

Thus, in the shock frame, the gas flows into the shock front at speed u_s and out at speed $u_s/4$. In an inertial frame in which the preshock gas is at rest, the gas is accelerated at the shock front to a speed $3u_s/4$.

The temperature rise at the discontinuity is given by

$$\frac{T_1}{T_0} = \frac{(p_1 + h^2 p_0) \, p_1}{(p_0 + h^2 p_1) \, p_0} \quad (2.54)$$

Thus, in the limit of $p_1 \gg p_0$, $T_1/T_0 \to (p_1/h^2 p_0)$. Unlike the compression ratio, the temperature ratio across the shock wave is unlimited.

The presence of a transverse magnetic field moderates the compression and the increase in temperature that occurs at the shock front. As we have already seen, if the magnetic field is sufficiently strong, the discontinuity can be suppressed altogether, in C-type shock waves.

2.3.2 The cooling flow of a J-type shock wave

If sufficiently hot, the compressed gas which flows out of a shock discontinuity is able to excite molecules, atoms and ions. These 'cooling' processes cause the temperature of the gas to fall whilst it continues to be compressed. By the time that the gas has cooled to its equilibrium (postshock) temperature, the total compression ratio, relative to the preshock gas, can be much greater than the maximum value of 4 at the shock discontinuity. The presence of a finite, transverse magnetic field limits the degree of compression of the gas.

2.3 The structure of interstellar shock waves

The temperature profile computed for a J-type shock wave with a speed $u_s = 25\,\mathrm{km\,s^{-1}}$, propagating into gas of (preshock) density $n_H = n(H) + n(H_2) + n(H^+) = 10^4\,\mathrm{cm^{-3}}$, in the absence of a magnetic field, is shown in Fig. 2.1. The independent variable in this figure is the flow time,

$$t = \int \frac{1}{u}\,dz$$

where z is the direction of flow and u is the flow speed, in the shock frame. In these calculations, the shock 'discontinuity' has a small but finite width, owing to artificial viscosity terms having been introduced into the conservation equations; these equations can then be integrated from the preshock through to the postshock, equilibrium gas.

The initial jump in temperature at the discontinuity is sufficient for collisional dissociation to take place, as may be seen from Fig. 2.1, where the abundances of H, H_2 and H^+, relative to n_H, are plotted. In fact, molecular hydrogen is rapidly collisionally dissociated in the hot gas behind the 'discontinuity', on a flow timescale of the order of 1 year. The main coolants of the gas are then atoms and ions, notably atomic oxygen, through its fine structure transitions at 63 and 147 μm (see Chapter 6). In the cooling flow, H_2 reforms (on grains), and the associated heating of the gas gives rise to the 'knee' in the temperature profile, apparent in the right-hand panel of Fig. 2.1. Finally, after approximately 500 years, the kinetic temperature reaches its postshock equilibrium value, of the order of 10 K. The total compression ratio in this case approaches 10^4: there is no magnetic field to moderate the compression of the gas.

Referring to equation (2.45), we see that, in the absence of a magnetic field, contributions to the energy flux arise from: (i) the kinetic energy of the flow; (ii) the thermal energy of the gas; (iii) the internal energy of the gas; (iv) radiative losses (incorporated in B). In the preshock gas, (i) dominates. Immediately behind the 'discontinuity', (ii) is the major term,

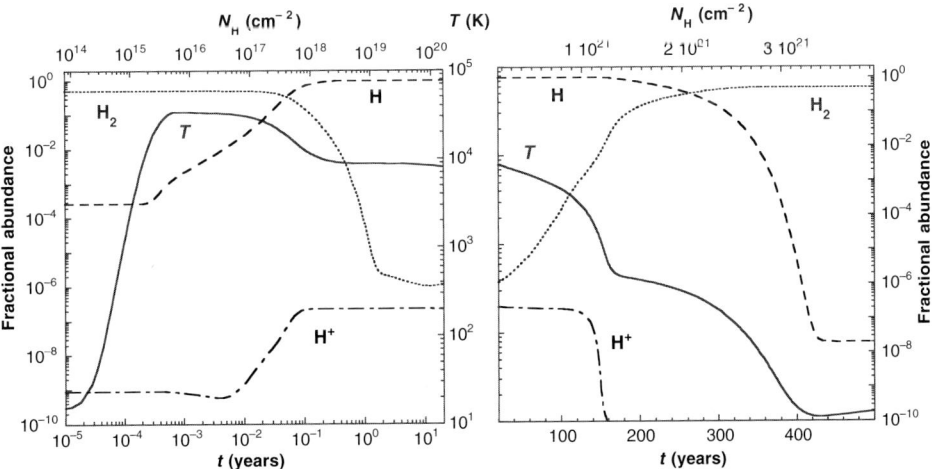

Figure 2.1 The temperature profile computed for a J-type shock wave with a speed $u_s = 25\,\mathrm{km\,s^{-1}}$, propagating into gas of (preshock) density $n_H = n(H) + n(H_2) + n(H^+) = 10^4\,\mathrm{cm^{-3}}$, in the absence of a magnetic field; N_H is the corresponding column density. The fractional abundances of H, H_2 and H^+ are also plotted. In the left-hand panel, the abscissa (the flow time) is on a logarithmic scale.

26   *Interstellar shocks and chemistry*

with some contribution from (iii) at the beginning of the cooling flow. Finally, radiative losses take over and the gas cools to its postshock, equilibrium state.

2.3.3 C-type shock waves

The interaction of the gas and the grains with the magnetic field is crucially important to the development of C-type shock waves. The field couples directly to the charged fluid and thence to the neutral fluid, which contains most of the mass, via collision processes (cf. Section 2.2 above). The strength of the coupling between the charged and neutral fluids depends on the degree of ionization of the gas and hence on the rates of chemical reactions which modify the degree of ionization within the shock wave. The coupling between the charged and neutral fluids also depends on the fraction of the grains that is (principally negatively) charged. Although the *number density* of the grains is much smaller than that of the gaseous ions, their *mass density* is, in dark clouds, much larger.

By way of illustration of the importance of chemical reactions in this context, Fig. 2.2 compares the fractional ionization of the gas and the thermal profiles computed for a C-type shock wave of speed $u_s = 10$ km s^{-1}, which propagates into gas of preshock density $n_H = 10^3$ cm^{-3} and in which the transverse magnetic field strength is $B_0 = 25$ μG; in one calculation, chemical reactions were neglected, and, in the other, they were included. As may be seen from this figure, the degree of ionization is modified considerably by the chemistry,

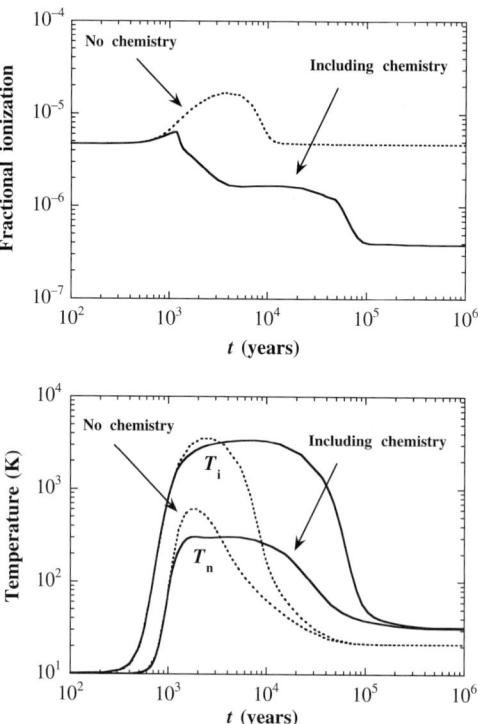

Figure 2.2 The fractional ionization of the gas and the thermal profiles in a C-type shock wave of speed $u_s = 10$ km s^{-1}, which propagates into gas of preshock density $n_H = 10^3$ cm^{-3} in which the transverse magnetic field strength $B_0 = 25$ μG.

2.3 The structure of interstellar shock waves

both within the shock wave and in the postshock, equilibrium gas. Endothermic reactions between atomic ions and H_2 are activated by the ion-neutral drift speed within the shock wave, and the molecular ions that are produced are able to recombine rapidly with electrons. Thus, the fractional ionization *falls* within the shock wave when the chemistry is included. On the other hand, when the chemistry is neglected, the fractional ionization increases, owing to the differential compression of the ions, before falling back to its equilibrium value in the postshock gas. To a lower degree of ionization corresponds a more extended shock wave, in which the neutral fluid has a lower maximum temperature.

Figure 2.2 also shows that the time of flow through a C-type shock wave, from the preshock to the postshock equilibrium gas is of the order of 10^5 years; this is very much greater than in a J-type shock wave of comparable speed. As the time to reach steady state cannot be less than the time of flow through the shock wave, it follows that C-type shock waves do not attain steady state under conditions, such as those in jets, in which the dynamical timescales are much less than 10^5 years.

In order to determine rigorously the structure of shock waves in their evolution to a steady state, a time-dependent MHD code is necessary. Such codes have been developed, but they are still restricted in the range and complexity of the physico-chemical processes that can be incorporated. An alternative approach, which provides an approximation to the time-dependent shock structure that is acceptable in the context of many applications, will be presented below.

When there is a disturbance that propagates at supersonic speed in a medium, a shock wave can be produced. Stellar winds, jets, turbulence, and collisions between interstellar clouds, for example, are all susceptible to generating shock waves. The concept of 'steady state' is relevant only if the mechanism responsible for producing the shock wave endures for at least the time required for the shock wave to attain its equilibrium state. Evanescent phenomena must, by their nature, be studied by means of an explicitly time-dependent model. The energy source that creates and maintains a shock wave may be compared with the 'piston' that can be used to generate shock waves in the laboratory. Let us denote the speed of propagation of the piston by u_p. We shall assume that u_p is constant and show, from considerations of the continuity of the flow, that the shock front propagates at a speed, u_s, which somewhat exceeds u_p.

As is customary, we apply the equation of continuity in the frame of the shock wave, i.e. the frame in which the shock front is at rest; this is achieved by subtracting the shock velocity, $\mathbf{u}_s$, from velocities in an inertial frame, usually taken to be the frame of the preshock gas. Referring to equation (2.47), we see that

$$u_1 = \frac{u_s}{(\rho_1/\rho_0)} \tag{2.55}$$

where ρ_0, ρ_1 denote the preshock and postshock gas density, respectively, and u_1 is the postshock flow speed; the ratio (ρ_1/ρ_0) is the compression factor. In the inertial frame, the preshock gas is at rest and the postshock gas flows at speed

$$u_s - \frac{u_s}{(\rho_1/\rho_0)}$$

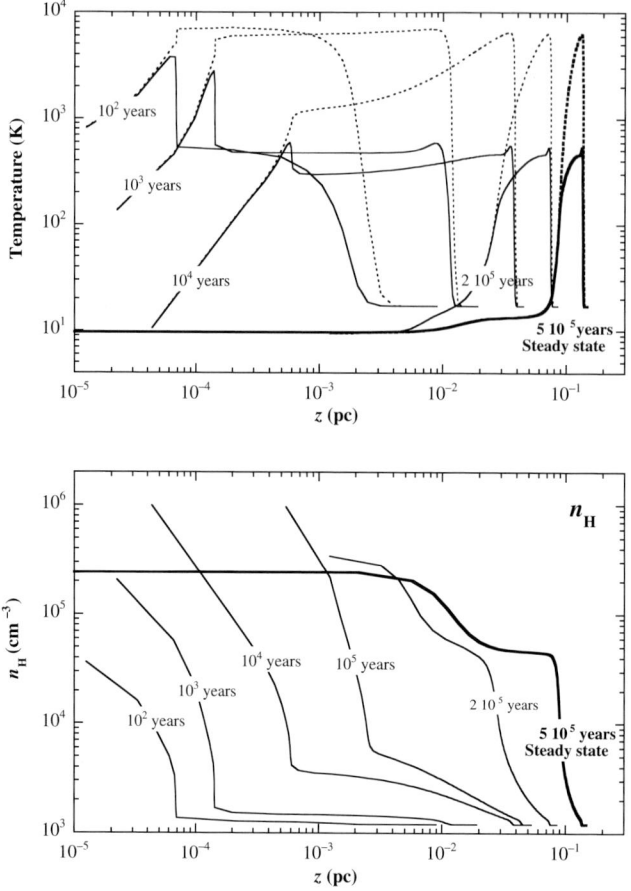

Figure 2.3 Temperature and density profiles as functions of position and time for a shock wave of speed $u_s = 10$ km s^{-1}, propagating into gas of density $n_H = 10^3$ cm^{-3} in which the transverse magnetic field strength $B_0 = 25$ μG. The shock wave advances from left to right until a stationary state is finally attained. The origin of the position coordinate, z, is at the 'piston'.

in the direction of propagation of the shock front. At the surface of the piston, the gas is moving with the speed of the piston, u_p, and hence

$$u_p = u_s \left[1 - \frac{1}{(\rho_p/\rho_0)} \right] \quad (2.56)$$

where ρ_p is the value of ρ_1 the surface of the piston. It follows from (2.56) that $u_s > u_p$.

If the shock is initially J-type, the compression factor at the discontinuity is $(\rho_1/\rho_0) = 4$, in the limit of large Mach numbers. Then, $u_s = 4u_p/3$, and the shock discontinuity moves away from the piston, with which it was initially in contact. In the cooling flow which develops between the discontinuity and the piston, the gas is compressed further and the speed of the shock front decreases towards that of the piston (which is assumed constant). By the time that the compression factor at the surface of the piston has become large $[(\rho_p/\rho_0) \gg 1]$, the piston

2.4 Shock waves in dark clouds

and the shock front are travelling at the same speed. From equation (2.56), it may be seen that the speed with which the shock front moves away from the piston, $u_s - u_p = u_s/(\rho_p/\rho_0)$, is also the speed of fluid flow (in the reference frame of the shock wave) at the surface of the piston.

If the transverse magnetic field strength is sufficiently large, a magnetic precursor develops upstream of the shock discontinuity and preheats the gas. As a consequence, the sound speed in the gas immediately upstream of the discontinuity increases and the Mach number falls, i.e. the shock discontinuity weakens. By the time that steady state is attained, the discontinuity may have disappeared altogether, in which case the structure has become pure C-type; this evolution is illustrated in Fig. 2.3. This figure shows also the shock front gradually separating from the 'piston' as time progresses; the speed of the shock front, relative to the piston, decreases as the compression factor increases.

The evolution of the shock wave, from initially pure J-type, to J-type with a magnetic precursor, seen in Fig. 2.3, finally to C-type, may be simulated by introducing a discontinuity in the flow at a point in the steady-state profile which is located increasingly downstream as time advances. The time is given by the time of flow of a fluid particle through the precursor to the discontinuity,

$$t = \int \frac{1}{u} \, dz$$

The steady-state structure of the shock wave 'unfolds' as time progresses. Comparisons with the results of explicitly time-dependent MHD calculations [32, 33] have shown that the evolution of the shock wave is satisfactorily described by means of the approximation outlined above.

2.4 Shock waves in dark clouds

The characteristics and spectroscopic signatures of J-type shock waves propagating in molecular media have been studied for many years. In the cooling flow behind the discontinuity, molecules, atoms and ions can be collisionally excited in the shock-heated gas. Rovibrational molecular transitions, fine structure and other 'forbidden' atomic and ionic transitions are emitted and, when detected, provide diagnostic information on the medium.

The physical state and chemical composition of the cooling flow depend on the shock speed, the transverse magnetic field strength in, and the density of, the preshock gas. With increasing shock speed, u_s, the maximum postshock temperature increases to values (of the order of 10^3 K) at which the collisional excitation of molecular hydrogen begins to be significant. As u_s increases further and the temperature reaches approximately 10^4 K, electronic excitation of atomic hydrogen occurs and, ultimately, H is collisionally ionized. The ultraviolet radiation produced by the radiative decay of the electronically excited states of H becomes sufficiently intense to pre-ionize the gas upstream of the shock discontinuity; the shock wave is then said to have a *radiative precursor*. If the transverse magnetic field is weak or absent, shock speeds $u_s \approx 25$ km s^{-1} are sufficient to cause almost complete dissociation of H$_2$ immediately downstream of the discontinuity.

In Fig. 2.4 the contributions of various atomic and molecular species to the cooling of a J-type shock wave of speed $u_s = 25$ km s^{-1}, preshock density $n_H = 10^4$ cm^{-3} and transverse magnetic field strength $B_0 = 0$ are illustrated. The collisional dissociation of molecular

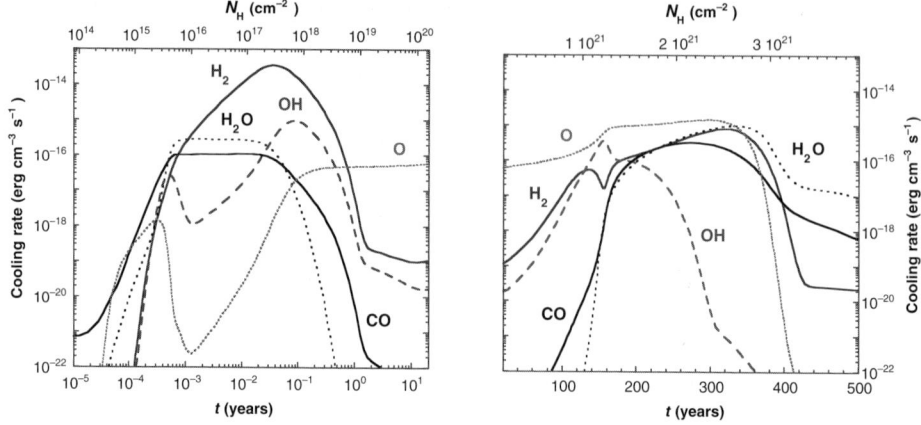

Figure 2.4 The rates of cooling by the principal coolants for a J-type shock wave with a speed $u_s = 25$ km s^{-1}, propagating into gas of (preshock) density $n_H = n(H) + n(H_2) + n(H^+) = 10^4$ cm^{-3}, in the absence of a magnetic field; N_H is the corresponding column density. In the left-hand panel, the abscissa (the flow time) is on a logarithmic scale. Note that 1 W $\equiv 10^7$ erg s^{-1}.

hydrogen releases atomic hydrogen into the hot gas, leading to the destruction of other molecular species. For example, the reactions

$$H_2O + H \rightarrow OH + H_2$$

and

$$OH + H \rightarrow O + H_2$$

return oxygen to atomic form. Through its fine structure transitions at 63 μm and 147 μm, O then becomes the principal coolant of the gas. As the gas cools down, first H_2 reforms, and then other molecules, such as H_2O and CO. The time required for the medium to reach its postshock equilibrium temperature is approximately 500 years. Thus, in dynamically young objects, even J-type shock waves may not have had sufficient time to reach steady state.

As the shock speed increases beyond $u_s = 25$ km s^{-1}, the extent of the cooling flow (and the time required for the gas to attain its postshock equilibrium temperature) begins to *decrease*. This reversal occurs because the degree of ionization of atomic hydrogen, and consequently the fractional electron abundance, increase rapidly with u_s above approximately 25 km s^{-1}. Cooling of the medium owing to electron collisional excitation of atomic hydrogen, principally the Ly α transition, then becomes important.

The populations of the rovibrational levels of H_2 do not respond instantaneously to the changes in temperature and density that occur at and behind the shock discontinuity. Indeed, as we have already seen, excitation of the internal degrees of freedom of H_2 is insignificant within the shock 'discontinuity', where the flow variables change adiabatically. In the cooling flow, immediately behind the discontinuity, the level populations respond to changes in the temperature and density on a timescale that, by definition, is comparable with the local cooling time (on which the temperature changes significantly), as H_2 is the principal coolant of the

2.4 Shock waves in dark clouds

gas. Under these conditions, it is essential to integrate the differential equations governing the H_2 level populations in parallel with the dynamical equations and the chemical rate equations.

In the presence of a transverse magnetic field of sufficient strength, an initially J-type shock wave develops a magnetic precursor and can ultimately become C-type. In order for a precursor to develop, the magnetosonic speed, c_m [equation (2.17)], must exceed the shock speed, u_s. As the physical conditions in the preshock gas determine the value of the magnetosonic speed, the requirement that $u_s < c_m$ sets an upper limit on the speed of shock waves that can become C-type.

The physical conditions in the preshock gas, notably the degree of ionization, depend on the rate of cosmic ray ionization of hydrogen, ζ. Most of the positive charge is associated with atomic and molecular ions in the gas. However, contributions to the negative charge arise not only from the free electrons but also from negatively charged grains and, possibly, from anions of polycyclic aromatic hydrocarbons (PAH). The distribution of the negative charge amongst free electrons, grains and PAH depends on the fractional abundance of the PAH molecules and on the rates of electron attachment and detachment reactions, as well as the rates of recombination of positive ions with electrons on the surfaces of negatively charged grains; all these parameters are subject to significant uncertainties.

In Fig. 2.5 the values of the magnetosonic speed computed for a range of densities of the preshock gas and of fractional abundances of PAH molecules are plotted; the cosmic ray ionization rate is $\zeta = 1 \times 10^{-17}$ s^{-1} and the transverse magnetic field strength $B_0 = [n_H]^{0.5}$, where n_H is in units of cm^{-3} and B_0 is in μG. The density dependence of B_0 is such as to

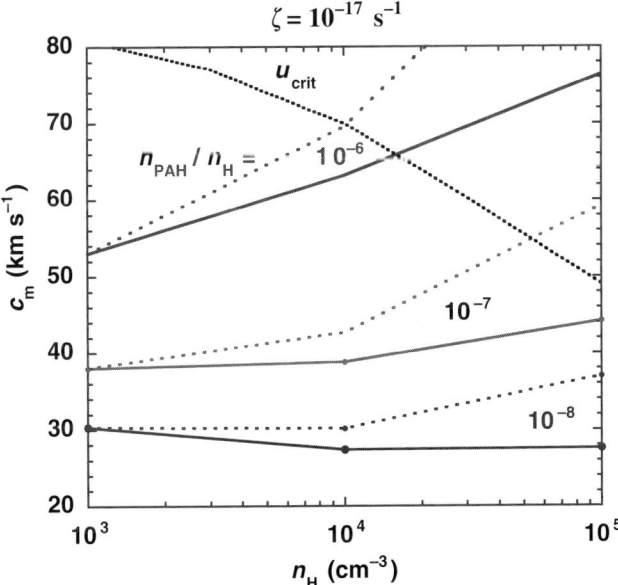

Figure 2.5 The full curves show the magnetosonic speed in preshock gas of density $10^3 \leq n_H \leq 10^5$ cm^{-3} and fractional abundance of PAH $10^{-8} \leq n_{PAH}/n_H \leq 10^{-6}$; the cosmic ray ionization rate is $\zeta = 1 \times 10^{-17}$ s^{-1} and the transverse magnetic field strength B_0 μG $= [n_H]^{0.5}$, where n_H is in units of cm^{-3}. The broken curves show the effect of correcting for collisional decoupling of the charged grains from the magnetic field.

ensure that the magnetic energy density in the preshock gas scales in proportion to the thermal energy density, at a given temperature.

Also plotted in Fig. 2.5 is the critical shock speed, u_{crit}, at which the degree of collisional dissociation of H_2, the principal coolant, becomes sufficient for a thermal runaway to occur. There results a sonic point in the flow (owing to the rapidly rising temperature and hence sound speed), and the shock becomes J-type. It may be seen from Fig. 2.5 that, for $n_{\mathrm{PAH}}/n_{\mathrm{H}} = 10^{-6}$ and $n_{\mathrm{H}} > 10^4$ cm^{-3}, the upper limit to the possible speed of a C-type shock wave is determined by the collisional dissociation of H_2, whereas, for $n_{\mathrm{H}} < 10^4$ cm^{-3}, it is determined by the magnetosonic speed in the preshock gas. The limit imposed by the magnetosonic speed becomes more stringent for $n_{\mathrm{PAH}}/n_{\mathrm{H}} < 10^{-6}$. Although still uncertain, the fraction of carbon which is believed to exist in 'very small grains' sets an upper limit to the fractional abundance of the PAH, namely $n_{\mathrm{PAH}}/n_{\mathrm{H}} \lesssim 10^{-6}$.

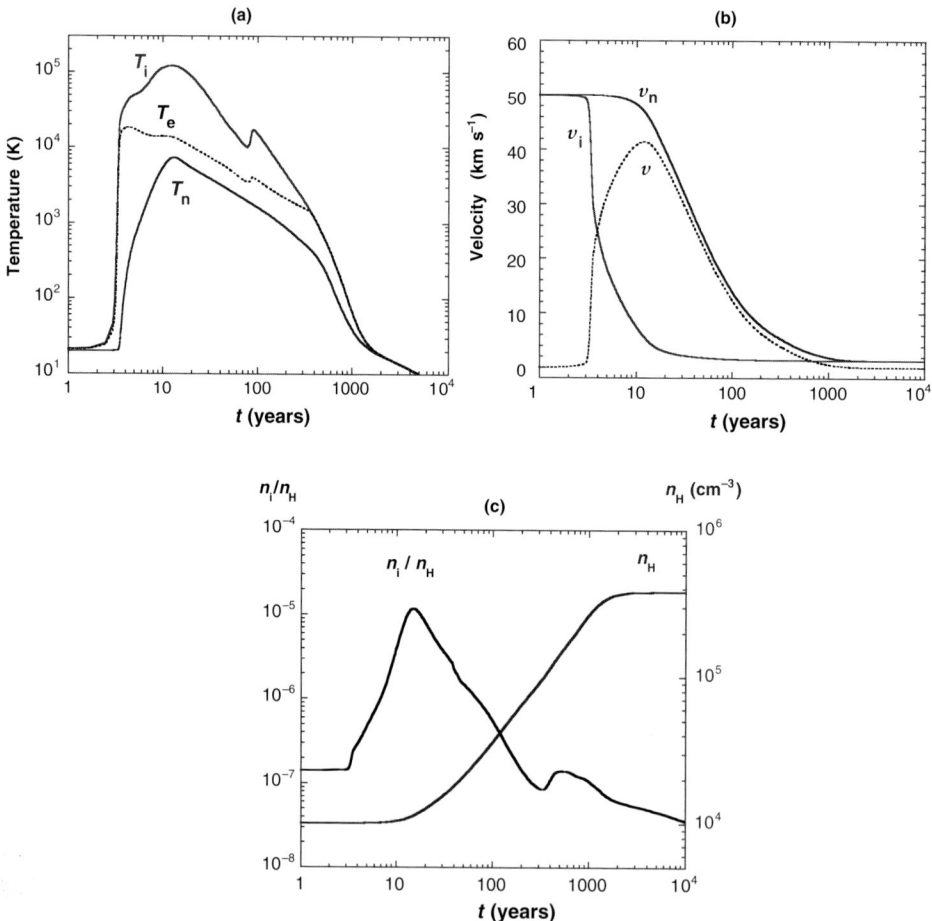

Figure 2.6 The steady-state profiles of (a) temperature, (b) velocity ($\Delta v = v_{\mathrm{n}} - v_{\mathrm{i}}$), and (c) density of the neutral and charged fluids, for an illustrative C-type shock wave of speed $u_{\mathrm{s}} = 50$ km s^{-1}, preshock density $n_{\mathrm{H}} = 10^4$ cm^{-3}, and transverse magnetic field strength $B_0 = 100$ μG.

2.5 Shock waves in diffuse clouds

The steady-state thermal, velocity and density profiles of the neutral and charged fluids for an illustrative C-type shock wave are shown in Fig. 2.6. These profiles display a number of characteristics of C-type shock structure: the initially rapid decoupling of the velocity of the charged fluid from that of the neutrals, followed by their recoupling on a timescale of the order of 10^3 years for the model shown; the initial increase in the fractional ionization, owing to the differential compression of the charged fluid, followed by a decrease (by approximately three orders of magnitude in the model shown) to its postshock value; the decoupling of the temperatures of the charged fluids from the temperature of the neutrals and, over a more restricted time range, the decoupling of the temperature of the ions from that of the electrons. The electron temperature lies between the temperatures of the ions and the neutrals: the much greater abundance of the neutrals compensates for the stronger thermal coupling of the electrons to the positive ions (which is mediated by the attractive coulomb force). Owing to the differential compression of the charged fluid and the initial rise in T_e, the rate of electron attachment to grains is sufficient to ensure that the grains are predominantly negatively charged within the shock wave. The rate equations that determine the grain charge must be solved in parallel with the equations that describe the dynamical structure of the shock wave, because the chemistry and the dynamics interact strongly.

2.5 Shock waves in diffuse clouds

The interstellar background ultraviolet radiation field permeates diffuse clouds and ionizes species with ionization potentials less than that of atomic hydrogen. Consequently, in diffuse gas, the most abundant ion is C^+. It has already been mentioned that endothermic reactions, such as

$$C^+ + H_2 \rightarrow CH^+ + H - 4640 \text{ K}$$

can become significant in shocked gas; this is true also of reactions such as

$$O + H_2 \rightarrow OH + H$$

which is endothermic and has a barrier, and of

$$OH + H_2 \rightarrow H_2O + H$$

which has a barrier. In a medium that is rich in molecular hydrogen, the reaction of C^+ with H_2 is followed rapidly by the exothermic hydrogenation reactions

$$CH^+ + H_2 \rightarrow CH_2^+ + H$$

and

$$CH_2^+ + H_2 \rightarrow CH_3^+ + H$$

beyond which the sequence proceeds much more slowly, either by radiative association

$$CH_3^+ + H_2 \rightarrow CH_5^+ + h\nu$$

or by the strongly endothermic reaction

$$CH_3^+ + H_2 \rightarrow CH_4^+ + H - 32\,500\text{ K}$$

All of the hydrocarbon ions which are formed undergo dissociative recombination reactions with electrons, such as

$$CH_n^+ + e^- \rightarrow CH_{n-1} + H$$
$$CH_n^+ + e^- \rightarrow CH_{n-2} + H_2$$
$$CH_n^+ + e^- \rightarrow CH_{n-2} + H + H$$

the net effect being the neutralization of C^+ ions in the gas.

The key reaction in the above hydrogenation cycle is $C^+(H_2, H)CH^+$, which is endothermic by 4640 K. Ambipolar diffusion in shock waves will drive this reaction once the relative kinetic energy of the ions and the neutrals is comparable with the endothermicity, i.e. once $m_{in}(u_i - u_n)^2/(2k_B) \approx 4640$ K, where $m_{in} = m_i m_n/(m_i + m_n)$ is the reduced mass of the C^+–H_2 pair. This relation implies that the relative drift speed, $(u_i - u_n)$, should be at least 6 or 7 km s^{-1}. Such speeds are readily attained in shock waves with speeds $u_s \gtrsim 10$ km s^{-1} in which the magnetic field strength in the preshock gas, B_0, is a few μG.

Photoreactions prevent ambipolar diffusion leading to the complete neutralization of the C^+ component of the ionized gas. The CH_n molecules, which are the products of the above cycle, are photodissociated

$$CH_n + h\nu \rightarrow CH_{n-1} + H$$

$$CH_n + h\nu \rightarrow CH_{n-2} + H_2$$

and the atomic carbon which is produced is then photoionized

$$C + h\nu \rightarrow C^+ + e^-$$

As a result of these reactions, C^+ ions are restored to the gas over a distance scale which is comparable with the MHD shock width. Thus, the C^+ density in such a shock wave first rises owing to the compression of the ionized gas, then falls as a result of ion–molecule reactions and dissociative recombination, and finally rises again as photoreactions take over. This behaviour is illustrated in Fig. 2.7. The pre- and postshock values of the density of C^+ ions are determined by the equilibrium between the rate of photoionization of C and the reverse process, namely, radiative recombination of C^+ with electrons.

It has already been mentioned that, in shock-heated gas, the chemistry of oxygen is initiated by the reaction $O(H_2, H)OH$, which has a barrier of 2980 K (which is larger than the endothermicity of approximately 960 K). The subsequent reaction, $OH(H_2, H)H_2O$, has a smaller barrier of 1490 K. Photodissociation of H_2O and OH eventually returns O to the gas. Reactions of C^+ and CH_n^+ with O and OH_n lead to the formation of CO^+ and H_nCO^+, which recombine dissociatively with electrons. The molecules which are so produced are ultimately photodissociated into C and O. The end result is the restoration of oxygen to its atomic form in the postshock gas. However, the facility with which water is produced in

2.5 Shock waves in diffuse clouds

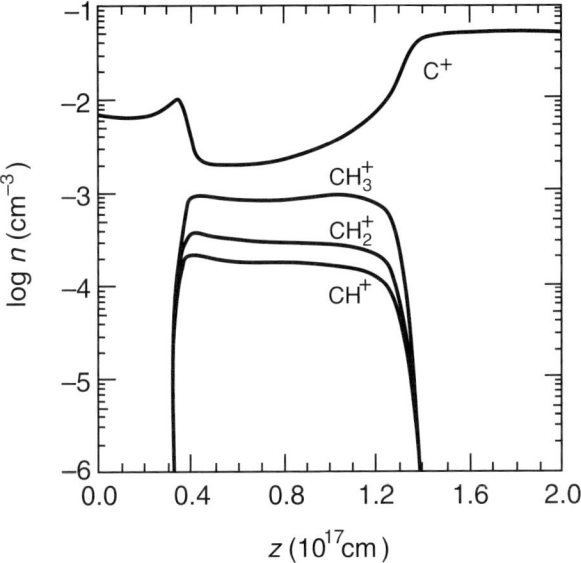

Figure 2.7 The densities of C-bearing ions through a C-type shock wave of speed 12 km s^{-1}, propagating into a diffuse interstellar cloud in which the preshock density $n_H = 20$ cm^{-3} and the transverse magnetic field strength $B_0 = 5$ μG.

shock waves, through the reactions O(H$_2$, H)OH and OH(H$_2$, H)H$_2$O, results in fractional abundances of H$_2$O, which exceed that observed in shocked molecular gas associated with IC443 (a supernova remnant) by the SWAS satellite [39]. Thus, although the transformation of atomic oxygen into water in shock waves was believed to be well understood, the models failed their first observational test. Other surprises of this type undoubtedly await us.

3

The primordial gas

3.1 Introduction

The interstellar medium has a structure and composition which have been modified by Galactic and stellar evolution over the lifetime of the Universe. While the gas remains rich in primordial H and He, it also contains small but significant amounts of C, N, O and other heavy elements, notably of the Fe-group, which have been produced by nucleosynthesis and then returned, in more or less violent events, to the interstellar medium. In addition to the gas, there is dust in this medium, which contributes only about 1% to the mass but which has important effects on the chemistry and the thermal balance (cf. Chapter 1). The primordial gas, on the other hand, contained no dust and was composed only of those elements that were produced in the primeval fireball – essentially hydrogen and helium, with trace amounts of deuterium, lithium and beryllium. Under these conditions, it is perhaps surprising that molecules existed and even more surprising that they should have played an important role in the evolution of the Universe. Nonetheless, this is believed to have been the case and, in this chapter, we explain why.

3.2 The governing equations

The cosmic background radiation field that we observe today has a black-body temperature of 2.73 K; it is a remnant of the 'big bang', which is believed to have occurred at the origin of the Universe. The energy density of the (black-body) radiation field is $U_r = aT_r^4$, where T_r is the radiation temperature and $a = 4\sigma/c$; σ is the Stefan–Boltzmann constant and c is the speed of light. The corresponding radiation pressure is $p_r = U_r/3$. In the case of an adiabatic process, the first law of thermodynamics tells us that

$$d(UV) + p dV = 0 \qquad (3.1)$$

where dV is the change in the volume V owing to the adiabatic expansion of the Universe. Hence,

$$d(T_r^4 V) + \frac{T_r^4}{3} dV = 0 \qquad (3.2)$$

or

$$\frac{1}{T_r}\frac{dT_r}{dt} + \frac{1}{3V}\frac{dV}{dt} = 0 \qquad (3.3)$$

3.2 The governing equations

where t is the time. If the volume is spherical, of radius R, then $V = 4\pi R^3/3$ and

$$\frac{1}{V}\frac{dV}{dt} = \frac{3}{R}\frac{dR}{dt} \quad (3.4)$$

Using equations (3.4) and (3.3), we find that, owing to the adiabatic expansion of the Universe, the temperature T_r of the radiation field fell according to the relation

$$-\frac{1}{T_r}\frac{dT_r}{dt} = \frac{1}{R}\frac{dR}{dt} \equiv H(t) \quad (3.5)$$

where $H(t)$ is the *Hubble parameter* at age t of the Universe. The present value of the Hubble parameter (the 'Hubble constant') is $H_0 \approx 70\,\text{km}\,\text{s}^{-1}\,\text{Mpc}^{-1}$. The inverse of the Hubble parameter is the timescale which characterizes the expansion of the Universe; its present value is $1/H_0 \approx 1.4 \times 10^{10}$ years.

The expansion of the Universe causes a redshift of radiation, owing to the Doppler effect. The ratio of the emitted to the observed frequency of the radiation is given by

$$\frac{\nu_e}{\nu_o} = \left(\frac{1 + v/c}{1 - v/c}\right)^{\frac{1}{2}} \quad (3.6)$$

where v is the speed of recession of the source and c is the speed of light. The redshift parameter, z, is defined as

$$z = \frac{\lambda_o - \lambda_e}{\lambda_e} \quad (3.7)$$

and hence the left-hand side of equation (3.6) may be written as $1 + z$; it may then be seen that, in the limit of $v \ll c$, $z \approx v/c$. The redshift is observed to increase linearly with distance to the source, and z is related to the radius, R, of the Universe at time t through

$$z = \frac{R_0}{R} - 1 \quad (3.8)$$

where R_0 is the current radius, when $z = 0$. After substituting for R and dR/dt from equation (3.8) and its derivative with respect to t, equation (3.5) may be integrated to give

$$T_r = T_0(1 + z) \quad (3.9)$$

where $T_0 = 2.73$ K is the current value of the radiation temperature.

Recombination of electrons and ions occurred at $z \approx 1000$, leaving a predominantly neutral medium composed of H and He, with trace amounts of D, Li and Be; there was no dust. At $z \approx 300$, the temperature of the gas, T_g, decoupled from that of the radiation field. If the subsequent expansion of the gas is adiabatic, equation (3.1) applies with $U_g = 3nk_B T_g/2$ and $p_g = 2U_g/3$, where n is the number density of the gas, whence

$$\frac{3}{2}d(nT_g V) + nT_g dV = 0 \quad (3.10)$$

The primordial gas

Then, using equation (3.4), we obtain

$$\frac{1}{T_g}\frac{dT_g}{dt} = -\frac{2}{R}\frac{dR}{dt} \tag{3.11}$$

as the total number of particles, nV, remains constant.

Comparing equations (3.5) and (3.11), we see that, beyond the point at which T_g decouples from T_r, T_g falls more rapidly than T_r as the Universe continues to expand. In practice, there remains some thermal coupling of the matter to the radiation field, owing to scattering processes, including Thomson scattering of the photons on free electrons; but the coupling is weak, and equation (3.11) provides a good approximation to the thermal evolution of the gas. The temperature of the gas would currently be $T_g \approx 10^{-2}$ K had equation (3.11) continued to hold. In fact, the formation of stars and galaxies has resulted in the gas temperature being much higher than this value and strongly dependent on conditions local to the source being observed.

Assuming that the total mass of the gas, ρV, is constant, we have that

$$\rho R^3 = \rho_0 R_0^3 \tag{3.12}$$

and hence, using (3.8),

$$\rho = \rho_0 (1+z)^3 \tag{3.13}$$

where ρ_0 is the current mass density.

The acceleration of particles located at radius R is determined by the gravitational attraction of the material within the sphere of radius R. Using Newton's second law of motion, it is readily shown that the rate of expansion of the Universe under the restraining action of gravity is determined by the equation

$$\frac{d^2 R}{dt^2} = -\frac{4\pi}{3} G \rho R \tag{3.14}$$

where G is the gravitational constant and R, ρ are the radius, mass density at time t. Using the identity

$$\frac{d^2 R}{dt^2} = \frac{1}{2}\frac{d}{dR}\left(\frac{dR}{dt}\right)^2 \tag{3.15}$$

and equation (3.12), we obtain

$$\frac{d}{dR}\left(\frac{dR}{dt}\right)^2 = -\frac{8\pi}{3} G \rho_0 \frac{R_0^3}{R^2} \tag{3.16}$$

In the Einstein–de Sitter model, the Universe expands monotonically and $dR/dt \to 0$ as $R \to \infty$. Adopting this limit, the integral of equation (3.16) is

$$\frac{1}{R^2}\left(\frac{dR}{dt}\right)^2 = H^2(t) = \frac{8\pi}{3} G \rho_0 \left(\frac{R_0}{R}\right)^3 \tag{3.17}$$

3.3 The role of molecules

At the present epoch, $H(t) = H_0$ and $R = R_0$. Hence, from equation (3.17),

$$\rho_0 = \frac{3H_0^2}{8\pi G} \tag{3.18}$$

This equation defines the *critical density* for which, according to the Einstein–de Sitter model, the Universe is just 'closed', i.e. the expansion speed tends asymptotically to zero. Taking $H_0 = 70 \,\text{km s}^{-1} \,\text{Mpc}^{-1}$, $\rho_0 = 9 \times 10^{-30} \,\text{g cm}^{-3}$.

It is believed that baryonic matter constitutes only a few per cent of the critical density. Most of the baryonic mass is in the form of H and He. Combining equations (3.13) and (3.18), we obtain

$$n(z) = \Omega_b \frac{3H_0^2(1 + (n_{\text{He}}/n_{\text{H}}))}{8\pi G m_{\text{H}}(1 + (4n_{\text{He}}/n_{\text{H}}))}(1+z)^3 \tag{3.19}$$

for the number density of baryons at redshift z; Ω_b is the ratio of the baryonic density to the critical density ($\Omega_b \approx 0.04$) and m_{H} is the mass of the hydrogen atom. Adopting $H_0 = 70 \,\text{km s}^{-1} \,\text{Mpc}^{-1}$ and $n_{\text{He}}/n_{\text{H}} = 0.08$, equation (3.19) yields

$$n(z) = 4.5 \times 10^{-6} \Omega_b (1+z)^3 \tag{3.20}$$

3.3 The role of molecules

A question that is relevant to all of us is: how did the first stars and galaxies form? If stars and galaxies had not formed, there would be no planets and life as we know it would not exist.

Let us suppose that the first stars formed by the gravitational contraction of matter, owing to inhomogeneities in the primordial gas. Gravitational contraction results in the production of heat, and the increasing gas pressure ultimately stops the contraction – unless the heat produced can be evacuated from the medium. Atomic hydrogen (and *a fortiori* helium) is an ineffective coolant of neutral gas: its first excited (electronic) state lies approximately $10 \,\text{eV}$ ($\approx 10^5$ K) above ground and is excited efficiently only by electron impact. The fact that the first stars formed successfully implies the presence of other coolants, effective at lower temperatures and excited by atomic impact; the only serious contenders are H_2 and, more surprisingly, HD.

The excited electronic states of H_2 and HD are no more accessible energetically than those of H. However, molecular hydrogen and its deuterated isotope, HD, have rovibrational states of lower energy, which can be excited at lower temperatures. Furthermore, because vibrational and rotational degrees of freedom are associated with the motion of the nuclei, rather than the electrons, these degrees of freedom can be excited effectively by atomic species of similar mass, principally H and He, which are the most abundant constituents of the post-recombination gas.

Simple considerations enable useful predictions to be made of the vibrational energy level structure of diatomic molecules such as H_2 and HD. In keeping with the Born–Oppenheimer approximation (cf. Chapter 4), we consider the nuclei (protons or deuterons) to move on the potential energy curve of the ground $X^1\Sigma_g^+$ electronic state. In the vicinity of its minimum, the potential has an approximately harmonic form, i.e.

$$v(r) = \frac{1}{2} K(r - r_0)^2 \tag{3.21}$$

where r_0 is the equilibrium distance and K is the 'force constant'. The nuclei oscillate (vibrate) in this well, between extrema on the potential walls. Their relative motion is that of a simple harmonic oscillator of reduced mass, μ; this problem is one of the few in quantum mechanics for which analytical solutions can be found. The eigenenergies are given by

$$E_v = \left(v + \frac{1}{2}\right)\hbar\omega_v \tag{3.22}$$

where $v = 0, 1, 2, \ldots$ is the vibrational quantum number and the angular frequency of vibration is

$$\omega_v = \left(\frac{K}{\mu}\right)^{\frac{1}{2}} \tag{3.23}$$

The 'force constant' is determined by the behaviour of the potential energy curve in the vicinity of r_0. Making a Taylor series expansion of $v(r)$ about $r = r_0$,

$$v(r) = v(r_0) + (r - r_0)\left[\frac{dv(r)}{d(r - r_0)}\right]_{r=r_0} + \frac{(r - r_0)^2}{2}\left[\frac{d^2v(r)}{d(r - r_0)^2}\right]_{r=r_0} + \cdots \tag{3.24}$$

and with $v(r_0) \equiv 0$ and $[dv(r)/d(r - r_0)]_{r=r_0} \equiv 0$ [r_0 is the location of the minimum of $v(r)$], we see from a comparison of equations (3.21) and (3.24) that

$$K = \left[\frac{d^2v(r)}{d(r - r_0)^2}\right]_{r=r_0} \tag{3.25}$$

The reduced mass is $\mu = m_H/2$ in the case of H_2, and $\mu = 2m_H/3$ in the case of HD, where m_H is the mass of the hydrogen atom. In H_2, $\hbar\omega_v = 4395$ cm$^{-1} \approx 6300$ K in the ground electronic state, and $r_0 = 0.742 \times 10^{-8}$ cm. As may be seen from equation (3.22), the vibrational energy levels are equally separated, by $\hbar\omega_v$, and they are energetically accessible at temperatures of a few thousand kelvin. Owing to its larger reduced mass, the vibrational spacing in HD is smaller, by a factor $[(2/3)/(1/2)]^{\frac{1}{2}} = 1.15$.

In practice, the potential energy curve deviates increasingly from harmonic form with increasing distance from r_0. As a consequence, the higher vibrational energy levels are closer together than is predicted by the simple harmonic oscillator model. There exist 15 bound vibrational states in the case of H_2.

In addition to their vibrational motion, the nuclei undergo end-over-end rotational motion. Once again, reasonable estimates of the energies involved may be derived from an elementary model. Classically, the magnitude, J, of the rotational angular momentum of the nuclei is given by

$$J = I\omega_J \tag{3.26}$$

3.3 The role of molecules

where I is the moment of inertia and ω_J is the angular speed of rotation. The associated rotational kinetic energy is

$$T = \frac{1}{2}I\omega_J^2 = \frac{J^2}{2I} \tag{3.27}$$

If the molecule rotates freely – which is the case when it is isolated – its total energy is equal to its rotational kinetic energy, E_J:

$$\frac{J^2}{2I} = E_J \tag{3.28}$$

The quantum mechanical equivalent of this classical expression is

$$\frac{1}{2I}J^2\psi = E_J\psi \tag{3.29}$$

where J^2 is the rotational angular momentum operator and ψ is the rotational wave function. The solutions of Schrödinger's equation (3.29) are the spherical harmonics

$$\psi(\theta,\phi) = Y_{JM}(\theta,\phi) \tag{3.30}$$

where (θ,ϕ) are the spherical polar angles that define the orientation of the internuclear axis with respect to a laboratory coordinate frame. The corresponding eigenenergies are

$$\begin{aligned} E_J &= J(J+1)\frac{\hbar^2}{2I} \\ &\equiv BJ(J+1) \end{aligned} \tag{3.31}$$

where $B = \hbar^2/(2I)$ is known as the 'rotational constant'; $B = 1/(2I)$ when atomic units are used. The rotational angular momentum quantum number $J = 0, 1, 2, \ldots$. The moment of inertia is given by

$$I = \mu r_0^2 \tag{3.32}$$

Taking $r_0 = 0.742 \times 10^{-8}$ cm for the ground state of H_2 and $\mu = m_H/2$ yields $B = 61.3$ cm^{-1}, which is close to the spectroscopic value of $B = 59.3$ cm^{-1}.

As may be deduced from equation (3.31), the spacing between successive rotational energy levels J and $J+1$ increases with J as $2(J+1)$. In fact, as H_2 exists in distinct ortho and para forms, in which J is odd and even, respectively, the spacing is $2(2J+3)$ in each form. There are 30 bound rotational levels in the lowest vibrational state ($v = 0$) of H_2.

The existence in H_2 of separate ortho and para forms arises from the identity of the two protons, which are fermions (particles with 1/2-integral spin). The Pauli exclusion principle, which applies to fermions but not to bosons (particles with integral spin), requires that the nuclear wave function should be *anti-symmetric* under exchange of the (identical) protons. The nuclear wave function is a product of the spatial part, equation (3.30), and a spin function. The spherical harmonic has symmetry $(-1)^J$ under exchange of the nuclei, which corresponds to the transformations $\theta \to \pi - \theta$ and $\phi \to \pi + \phi$. The nuclear spin functions are symmetric under proton exchange in the ortho state (total nuclear spin $I = 1$) and asymmetric in the para state ($I = 0$). Thus, ortho states must associate with *odd* values of J, and para states with *even* values of J, in order to satisfy the exclusion principle. This dichotomy does not occur in HD, which is heteronuclear. Accordingly, the first rotationally excited state of HD, $J = 1$, can be reached from the ground state, $J = 0$, in *non*-reactive collisions. Furthermore, the rotational constant of HD is only 3/4 that of H_2, owing to the larger reduced mass of HD. As a consequence of these two effects, the energy required to excite the first rotational transition in HD ($J = 0 \to J' = 1$: 128.4 K) is approximately four times smaller than in the case of H_2 ($J = 0 \to J' = 2$: 509.9 K). Given that the rate coefficient for collisional excitation decreases exponentially with increasing energy

$$\langle \sigma v \rangle_{J' \leftarrow J} \propto \exp\left[\frac{-(E_{J'} - E_J)}{k_B T}\right]$$

it is clear that HD is a much more efficient coolant than H_2 at low temperatures, $T \lesssim 100$ K (cf. Chapter 5). Thus, even allowing for the small value of the elemental abundance ratio, $n_D/n_H \approx 4 \times 10^{-5}$, in the primordial medium, the contribution of HD to the cooling of low-temperature gas is significant. The importance of HD as a coolant is enhanced further by its chemical fractionation, considered in the following section.

Pursuing the same logic, one might consider whether molecules such as lithium hydride, LiH, which forms through the radiative association of Li and H, might contribute significantly to the cooling of the medium. LiH has a large dipole moment, and the separation of the $J = 1$ and $J = 0$ rotational energy levels is only about 8 K, owing to the reduced mass effect. However, it turns out that the low abundance of ^{7}Li in the primordial gas ($n_{Li}/n_H = 2 \times 10^{-10}$) and the slow rate of its radiative association with H [40] rule out LiH as an effective coolant.

3.4 Chemistry

In the interstellar medium, H_2 and HD form predominantly on the surfaces of grains, as discussed in Chapter 1. However, there were no grains in the primordial medium, and these molecules, if present, must have formed by other mechanisms. The two chemical routes to H_2 in the primordial gas are believed to have been

$$H + e^- \to H^- + h\nu \qquad (3.33)$$

followed by

$$H^- + H \to H_2 + e^- \qquad (3.34)$$

and

$$H + H^+ \to H_2^+ + h\nu \qquad (3.35)$$

followed by

$$H_2^+ + H \to H_2 + H^+ \qquad (3.36)$$

In both cases, the first step [attachment in equation (3.33), association in equation (3.35)] is radiative and hence slow. Indeed, this is one of the reasons that these reactions are insignificant sources of H_2 in the interstellar medium. Nonetheless, reactions (3.33)–(3.36) resulted in a

3.4 Chemistry

small but significant amount of H$_2$ [$n(H_2)/n(H) \approx 10^{-6}$] being present in the primordial gas by the time at which *freeze-out* occurred, and beyond which there was no further change in its fractional abundance ($z \lesssim 100$).

Deuteration of H$_2$ occurs in the reaction

$$D^+ + H_2 \rightleftharpoons H^+ + HD \tag{3.37}$$

which is exoergic in the forward (left-to-right) direction by 464 K when ground state reactants and products are involved. Consequently, the backward reaction is inhibited when the gas temperature falls below about 500 K, and the abundance of HD, relative to H$_2$, becomes enhanced relative to the elemental abundance ratio, $n_D/n_H = 4 \times 10^{-5}$; this process is known as *chemical fractionation*.

The rate coefficients, k_f and k_r, for the forward and reverse reactions (3.37), respectively, are related through detailed balance

$$\omega_f k_f = \omega_r k_r \exp\left(\frac{\Delta E}{k_B T_g}\right) \tag{3.38}$$

where ω denotes a statistical weight and f, r correspond to the left-to-right and right-to-left directions of reaction (3.37), respectively. As already mentioned, if the reactants and products are in their ground states, $\Delta E = 464$ K. An alternative form of equation (3.38), which is more usual in the chemistry literature, involves the enthalpy change (ΔH) and the entropy change (ΔS) in the reaction,

$$k_f = k_r \exp\left(\frac{\Delta S}{R}\right) \exp\left(\frac{-\Delta H}{RT_g}\right) \tag{3.39}$$

where R is the universal gas constant. Measurements show that $\Delta S/R = \ln(2.1)$ and k_f is close to the Langevin value of 2.1×10^{-9}.

Comparing equations (3.38) and (3.39), we see that

$$\Delta H/R = -\Delta E/k_B$$

or

$$\Delta H = -N_A \Delta E$$

where $N_A = 6.022 \times 10^{23}$ mol^{-1} is the Avogadro number. Similarly,

$$\Delta S = R \ln\left(\frac{\omega_r}{\omega_f}\right)$$

When reaction (3.37) has attained equilibrium, we have that

$$\frac{n(HD)}{n(H_2)} = \frac{n(D^+)}{n(H^+)} \frac{k_f}{k_r}$$

$$= \frac{n(D^+)}{n(H^+)} 2.1 \exp(464/T_g) \tag{3.40}$$

44 The primordial gas

As $n(D^+)/n(H^+)$ reflects the elemental abundance ratio, $n_D/n_H = 4 \times 10^{-5}$, the ratio $n(HD)/n(H_2) >> n_D/n_H$ when $T_g << 500$ K. Calculations show that, subsequent to freeze-out, the ratio $n(HD)/n(H_2) \approx 10^{-3}$, i.e. 25 times the elemental abundance ratio. The combination of chemical fractionation and the greater accessibility of the rotationally excited states of HD ensures that the importance of HD as a coolant of the primordial gas is comparable to that of H_2.

The analysis above applies when the reactants and products are in their ground states. When the molecules are in excited states, the exoergicity of the forward reaction (3.37) is modified. In practice, the rotational levels of HD, which are connected by electric dipole transitions, and even those of H_2, which has an electric quadrupole spectrum, are coupled to the background black-body radiation field. In the case of HD,

$$\frac{n_J}{n_0} = (2J+1)\exp\left(\frac{-E_J}{k_B T_r}\right) \tag{3.41}$$

where E_J is the energy of level J relative to the rotational ground state, $J = 0$, and T_r is the radiation temperature. The level populations of the ortho and para forms of H_2 thermalize *separately* to the background radiation temperature. On the other hand, the ortho : para H_2 ratio (the ratio of the sums of the populations of the odd-J and the even-J levels) does *not* thermalize to T_r, as the probabilities of radiative transitions between the two forms of H_2 are extremely small. The ortho and para forms interconnect through proton exchange in reactions with H^+ and H; but these reactions are not necessarily sufficiently rapid to ensure thermalization at the gas temperature, T_g. It follows that, when account is taken of the level populations of reactants and products, the rate of the deuteration reaction (3.37) depends on the ortho : para H_2 ratio and hence on the rates of proton exchange reactions taking place in the medium.

3.5 Gravitational collapse

In the uniformly expanding primordial gas, the contributions of both H_2 and HD to the thermal balance of the gas are, in fact, small, and the expansion remains adiabatic, to a good approximation. Furthermore, H_2 and HD *heat* the gas, rather than cool it, because the gas temperature $T_g \leq T_r$: energy is absorbed from the background radiation field and added to the kinetic energy of the gas by the process of collisional de-excitation. The net rate of heating is given by

$$\Gamma - \Lambda = \sum_{i>j}(\Gamma_{ij} - \Lambda_{ij}) \tag{3.42}$$

where

$$\Gamma_{ij} = \sum_p n^{(p)} q^{(p)}_{j \leftarrow i}(T_g) n_i (E_i - E_j) \tag{3.43}$$

and

$$\Lambda_{ij} = \sum_p n^{(p)} q^{(p)}_{i \leftarrow j}(T_g) n_j (E_i - E_j) \tag{3.44}$$

3.5 Gravitational collapse

where $q^{(p)}_{j \leftarrow i}(T_g)$ is the rate coefficient for collisional de-excitation $j \leftarrow i$ by perturber p – essentially by H, He and H$^+$; $q^{(p)}_{i \leftarrow j}(T_g)$ is the corresponding rate coefficient for collisional excitation. The net rate of heating, $\Gamma - \Lambda$, is zero when $T_g = T_r$ and the rotational level populations are thermalized at temperature T_r; but when $T_g < T_r$, which is the case for $z \lesssim 300$, $\Gamma - \Lambda > 0$.

At some stage in the expansion of the Universe, inhomogeneities developed in the primordial gas, as observations with the *Cosmic Background Explorer* (COBE) satellite have shown. The presence of inhomogeneities in the medium is essential for gravitational collapse to occur. Once gravitational collapse is under way, the kinetic temperature decouples from that of the radiation field, and it is the collisional excitation of H$_2$ and HD, followed by the emission of electric quadrupole and electric dipole radiation, respectively, which moderates the increase in the kinetic temperature of the gas.

The rates of collisional excitation of H$_2$ and HD depend on the abundances of the perturbers and on the corresponding rate coefficients. At kinetic temperatures $T_g \lesssim 1000$ K, the rate coefficients for rotational excitation of H$_2$ by He and H$_2$ are larger than for excitation by H. On the other hand, the elemental abundance of helium, relative to hydrogen, is only $n_{He}/n_H = 0.08$. Furthermore, until the process of three-body recombination of H to form H$_2$ begins to take effect, for gas densities $n_H \gtrsim 10^{10}$ cm^{-3}, $n(H_2) \ll n(H)$. Thus, He and H are the principal contributors to the rotational excitation and cooling of H$_2$, up to and somewhat beyond the point ($n_H \approx 10^8$ cm^{-3}) at which the rotational transitions begin to become optically thick. Vibrational excitation is dominated by H, not only because of its higher abundance but also because the rate coefficients for vibrational excitation by H are much larger than for excitation by He or H$_2$.

The electric quadrupole transition probabilities of H$_2$ are very small, increasing from approximately 3×10^{-11} s^{-1} for the 0-0 S(0) transition to values of the order of 10^{-6} s^{-1} for the high rotational and for rovibrational transitions. (Electric quadrupole transition probabilities $A \propto \nu^5$, where ν is the frequency of the transition: see Chapter 10.) Consequently, as the gas density increases, the populations of the lowest energy levels thermalize first.

At densities n_H much less than the critical density for thermalization, at which the rates of radiative and collisional de-excitation are equal, the rate of cooling is $\propto n_H^2$; collisional excitation is followed by radiative decay, and the photon escapes. Above the critical density, the level populations approach a Boltzmann distribution; the excited level populations and hence the rate of cooling is $\propto n_H$. The *cooling function*, which is defined as the rate of cooling of the gas per coolant molecule, has been computed for H$_2$ and HD by a number of workers; the results are in good agreement for kinetic temperatures $T_g \lesssim 1000$ K. However, the cooling function is an imperfect guide to the thermal evolution of primordial material which is undergoing collapse. Fixed relative abundances of the perturbers are assumed when evaluating the cooling function, whereas, in fact, the relative abundances vary as the collapse proceeds. Similarly, a constant value of the ortho : para H$_2$ ratio is adopted, whereas this ratio is non-thermal and time-dependent, in general. Finally, the density dependence of the populations of the rotational energy levels is often treated crudely, by means of computations in 'low' and 'high' density limits. In practice, the only way to allow properly for the cooling by H$_2$ and HD is through time-dependent calculations of the populations of their energy levels, with allowance for all significant collisional and radiative processes, including the coupling to the cosmic background radiation field.

46   The primordial gas

The equation governing the free-fall gravitational collapse of a condensation of gas, assumed spherical and of radius r, is

$$\frac{1}{r}\frac{dr}{dt} \equiv \frac{1}{x}\frac{dx}{dt} = -\frac{\pi}{2t_{\mathrm{ff}}x}\left(\frac{1}{x}-1\right)^{\frac{1}{2}} \quad (3.45)$$

where $x = r/a \leq 1$; a is the initial radius of the condensation. The timescale for free-fall collapse is [41]

$$t_{\mathrm{ff}} = \left[\frac{3\pi}{32G\rho(0)}\right]^{\frac{1}{2}} \quad (3.46)$$

where $\rho(0)$ is the initial mass of gas per unit volume, i.e. the density when $t = 0$ and $r = a$. Equation (3.45) may be derived by integrating equation (3.16) and applying the boundary condition that $dr/dt \to 0$ in the limit of $r \to a$. Owing to the (assumed spherically symmetric) collapse, the mass density of the gas, $\rho(t)$, varies with time t according to

$$\frac{1}{\rho}\frac{d\rho}{dt} = -\frac{3}{x}\frac{dx}{dt} \quad (3.47)$$

whereas the number density, $n(X)$, of the species X is determined by

$$\frac{1}{n(X)}\frac{dn(X)}{dt} = -\frac{3}{x}\frac{dx}{dt} + \frac{\mathcal{N}(X)}{n(X)} \quad (3.48)$$

where $\mathcal{N}(X)$ is the rate per unit volume at which the species X is created in chemical reactions.

In the collapsing condensation, the kinetic temperature of the gas is determined by the heating and cooling processes, Γ and Λ, the heating associated with the compression due to the collapse, the endo- and exo-thermicities of chemical reactions, and by the variation of the internal energy, U, of the molecules, principally of H_2. The kinetic temperature, T_g, varies with time according to

$$\frac{3}{2}nk_B\frac{dT_g}{dt} = \Gamma - \Lambda - \frac{3}{2}nk_BT_g\frac{\mathcal{N}}{n}$$
$$-\frac{d(nU)}{dt} + n(U + k_BT_g)\frac{1}{\rho}\frac{d\rho}{dt} \quad (3.49)$$

where $n = \sum_X n(X)$ is the number density of all species in the gas and $\mathcal{N} = \sum_X \mathcal{N}(X)$. The adiabatic limit is obtained on taking Γ, Λ and $\mathcal{N}$ equal to zero in equation (3.49). In this limit, the solution of (3.49) has the anticipated form, $T/\rho^{\gamma-1} =$ constant, where γ is the ratio of the specific heats at constant pressure and volume. For a monatomic gas, $\gamma - 1 = 2/3$, whereas, for a diatomic gas, $\gamma - 1 = 2/5$; these exponents are obtained on setting the internal energy $U = k_BT$ in equation (3.49), corresponding to $k_BT/2$ per rotational degree of freedom, in thermodynamic equilibrium.

The starting conditions for the free-fall collapse are determined by the initial value of the redshift parameter, z, and the baryon density, Ω_b, relative to the critical density. As the Universe expands adiabatically, z decreases and the radiation and matter temperatures decrease according to equations (3.5) and (3.11) above. The first phase of the free-fall collapse

3.5 Gravitational collapse

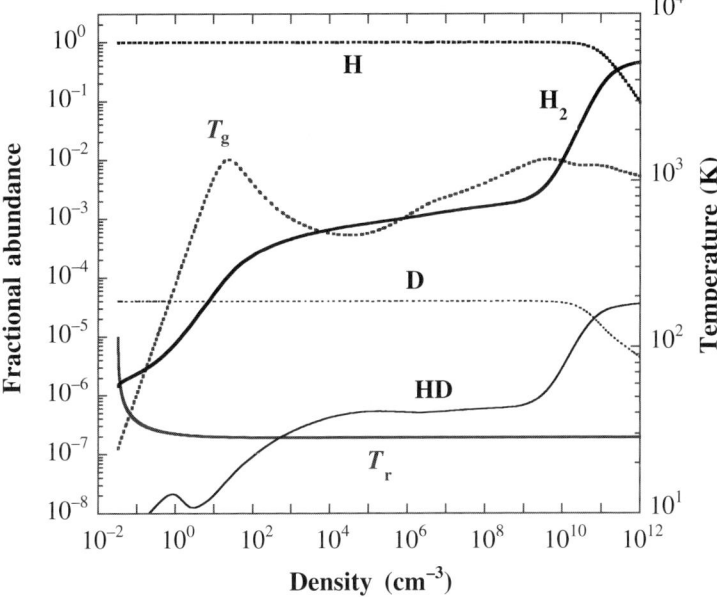

Figure 3.1 The thermal and chemical profiles of a collapsing condensation of gas: T_g is the temperature of matter, T_r is the temperature of the radiation field. The fractional abundances of the principal species are also plotted, relative to $n_H = n(H) + 2n(H_2) + n(HD)$.

is the reverse process of adiabatic expansion, i.e. adiabatic contraction. Thus, if the free-fall collapse commences at a smaller value of z, there results a somewhat longer duration of the adiabatic contraction phase of the collapse, and the same evolutionary track is followed subsequently. It follows that the profiles of kinetic temperature and density which are followed during the collapse are almost independent of the initial value of z.

The thermal and chemical profiles of a collapsing condensation of gas are illustrated in Fig. 3.1. The gas temperature, T_g, first increases adiabatically, owing to the contraction, until the density becomes sufficient for collisional cooling to reverse the rise in temperature; T_g then falls to a minimum when the population densities of the rotational levels of H_2 begin to thermalize. At this point, the contributions of H_2 and HD to the cooling of the gas are approximately equal. At densities approaching 10^{10} cm^{-3}, three-body association reactions begin to convert H into H_2 and D into HD. Although these reactions release energy and become a major source of heating of the gas, this increase in the heating rate is compensated by the simultaneous rise in the fractional abundance and hence the rate of cooling by H_2. At this stage in the contraction, H_2 is a more important coolant of the gas than HD: the abundance ratio $n(HD)/n(H_2) \to 2n_D/n_H = 8 \times 10^{-5}$ in the limit where practically all of the hydrogen and the deuterium are in H_2 and HD, respectively; in other words, the HD : H_2 molecular abundance ratio is enhanced by no more than a factor 2 relative to the D : H elemental abundance ratio.

The results in Fig. 3.1 were derived assuming that the homologous, free-fall collapse described by equation (3.45) remains valid. In practice, the speed of collapse comes to exceed the adiabatic sound speed beyond the initial maximum in T_g. Under such conditions, shock waves are likely to develop in the collapsing gas, causing local heating and compression; the resulting filaments of compressed gas will subsequently fragment. This process of fragmentation has important consequences for the initial mass function (IMF), i.e. the mass spectrum, of the first objects that condense from the primordial gas.

The evolution of the filaments that are formed during the process of collapse may be studied by means of the virial equation. Assuming a cylindrical geometry, this equation may be written as

$$\frac{1}{2}\frac{d^2 I}{dt^2} = 2T + \frac{4}{3}U + 2M - Gm^2 \quad (3.50)$$

as shown in the remarkable paper of Chandrasekhar and Fermi [42], where $I = mr^2/2$ is the moment of inertia per unit length of a cylinder of radius r and mass per unit length $m = \pi r^2 \rho$, $T = m(dr/dt)^2/4$ is the kinetic energy, $U = \pi r^2(3/2)nk_B T_g$ is the thermal energy, and $M = \pi r^2[B^2/(8\pi)]$ is the magnetic energy; B is the magnetic induction. If we neglect the magnetic energy, about which little is known, in any case, then equation (3.50) yields a necessary condition for collapse to occur from rest, namely

$$\frac{4}{3}U - Gm^2 < 0$$

Using the definitions of U and m, and setting $\rho = n\mu$, where μ is the mean molecular weight of the gas, this condition becomes

$$\frac{2k_B T_g}{\mu} < Gm$$

or

$$m > \frac{2k_B T_g}{\mu G}$$

which is the minimum mass per unit length of the cylinder that is required for the cylinder to collapse.

Cylindrical filaments that are collapsing under the effect of self-gravity tend to fragment when the dynamical timescale, $\rho/(d\rho/dt)$, becomes greater than the fragmentation timescale, which is given by

$$t_f \approx \frac{3}{(4\pi G\rho)^{\frac{1}{2}}}$$

The masses of the filaments formed are approximately equal to $2\pi r_f m$, where r_f is the radius of the cylinder when fragmentation occurs. The IMF depends on the initial conditions and, in particular, on the fractional abundance of H_2. Both H_2 and HD played key roles, as coolants of the gas, in determining the IMF of the first generation of stars.

In practice, sophisticated hydrodynamic calculations are necessary in order to attempt to establish the IMF quantitatively. The mass distribution of the initial stars is crucial to the subsequent evolution of the medium. The radiation field changes when stars form, and the composition of the medium is modified as stars evolve and return material to the medium. These evolutionary processes depend on the initial mass function.

4

The rotational excitation of molecules

4.1 Introduction

The quantum theory of molecular collisions has been extensively developed over the last three decades. As in many branches of theoretical science, the growth of this subject has been closely linked with the advances in computer technology. Powerful numerical techniques have been developed for solving Schrödinger's equation, which are well adapted to low energy, molecular collision problems, at various levels of approximation. A basic reference text in this context is *Atomic, Molecular and Optical Physics Handbook* [43]. The complexity of the problems that can be tackled, and the accuracy of the results that can be obtained, continue to be determined by the available computing power.

Any proper discussion of molecular collision processes involves the concept of the potential energy curve or surface. This concept drives from the Born–Oppenheimer approximation, to which we first turn.

4.2 The Born–Oppenheimer approximation

For the sake of simplicity, when discussing the basic concepts, we consider the collision between a one-electron atom, A, and a fully-stripped ion, B. The theory which pertains to this illustrative case can be generalized to collisions between many-electron atoms or to collisions between molecules.

When studying a collision problem, we are interested in the relative motions of the particles involved, and not in the motion of the centre of mass (*barycentre*) of the colliding system. The velocity of the centre of mass remains constant and is irrelevant to the scattering processes to which we progress below. The position of the centre of mass, relative to a space-fixed or laboratory reference frame, will be denoted by $\mathbf{R}_C$ and the coordinates of the centre of mass of A and of B by $\mathbf{R}_A$ and $\mathbf{R}_B$, respectively. $\mathbf{R}$ is the vector connecting the centre of mass of A to B and is taken to be directed from A to B; see Fig. 4.1. The position of the centre of mass of the system is defined such that

$$\mathbf{R} = \mathbf{R}_{AC} + \mathbf{R}_{CB}$$

$$= \frac{m_B}{m_A + m_B}\mathbf{R} + \frac{m_A}{m_A + m_B}\mathbf{R} \qquad (4.1)$$

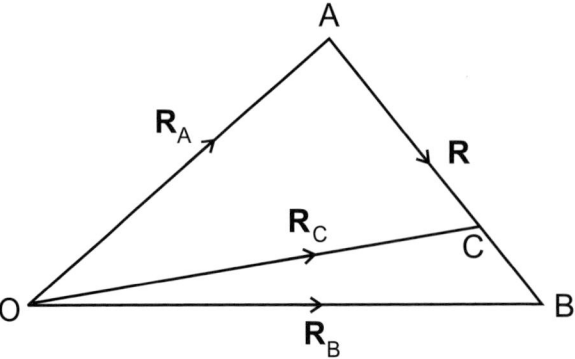

Figure 4.1 Defining the space-fixed or laboratory coordinates of an atom, A, and a fully-stripped ion, B, and of their centre of mass, C; the relative coordinate, **R**, is directed from A to B.

and the momentum of the system is

$$m_A \dot{\mathbf{R}}_A + m_B \dot{\mathbf{R}}_B$$
$$= m_A (\dot{\mathbf{R}}_C - \dot{\mathbf{R}}_{AC}) + m_B (\dot{\mathbf{R}}_C + \dot{\mathbf{R}}_{CB})$$
$$= (m_A + m_B) \dot{\mathbf{R}}_C \tag{4.2}$$

where use has been made of equation (4.1). Thus, the momentum of the system may be considered as being due to the total mass, $(m_A + m_B)$, located at the centre of mass.

Consider now the kinetic energy of the system,

$$\frac{1}{2} m_A \dot{\mathbf{R}}_A^2 + \frac{1}{2} m_B \dot{\mathbf{R}}_B^2$$
$$= \frac{1}{2} m_A (\dot{\mathbf{R}}_C - \dot{\mathbf{R}}_{AC})^2 + \frac{1}{2} m_B (\dot{\mathbf{R}}_C + \dot{\mathbf{R}}_{CB})^2$$
$$= \frac{1}{2} (m_A + m_B) \dot{\mathbf{R}}_C^2 + \frac{1}{2} \mu \dot{\mathbf{R}}^2 \tag{4.3}$$

where $\mu = m_A m_B/(m_A + m_B)$ is the *reduced mass* of the system. Thus, the kinetic energy comprises contributions from the total mass, moving with the velocity of the centre of mass, $\dot{\mathbf{R}}_C$, and from the reduced mass, moving with the relative velocity, $\dot{\mathbf{R}}$, of A and B. When the system is isolated, as we assume, the velocity of the centre of mass is constant and may be removed by a change of inertial frame; there remains the relative kinetic energy, which is available for exciting the internal degrees of freedom of the colliding system. Thus, we may consider the atom A to move with reduced mass, μ, relative to a fixed centre of force, B.

When discussing the interactions between atoms and molecules, it is often convenient to use the atomic system of units, in which $e = m_e = \hbar = 1$. In these units, the hamiltonian of the system AB may be written

$$H(\mathbf{x}, \mathbf{R}) = H_A(\mathbf{x}, \mathbf{R}) - \frac{\nabla_R^2}{2\mu} + V(\mathbf{x}, \mathbf{R}) \tag{4.4}$$

4.2 The Born–Oppenheimer approximation

where **x** is the position vector of the electron with respect to the centre of mass of A and B, $H_A(\mathbf{x}, \mathbf{R})$ represents the electronic hamiltonian of the atom A and $V(\mathbf{x}, \mathbf{R})$ is the potential of interaction between A and B; $-\nabla_R^2/(2\mu)$ is the relative kinetic energy operator, and ∇^2 the Laplacian operator ($\nabla_R^2 = \frac{\partial^2}{\partial X^2} + \frac{\partial^2}{\partial Y^2} + \frac{\partial^2}{\partial Z^2}$). Our task is to solve the Schrödinger equation

$$H\Psi = E\Psi \tag{4.5}$$

where E is the total barycentric energy of the colliding system, which is the kinetic energy of relative motion of A and B at infinite separation. The wave function is $\Psi = \Psi(\mathbf{x}, \mathbf{R})$.

Let us write the wave function in the following form, which retains generality but hints at the separation of the dependences on the electronic coordinates, **x**, and the relative coordinates, **R**:

$$\Psi(\mathbf{x}, \mathbf{R}) = \sum_i F_i(\mathbf{R}) \phi_i(\mathbf{x}, \mathbf{R}) \tag{4.6}$$

For fixed R, equation (4.6) is an expansion of the wave function in terms of the solutions of the wave equation

$$[H_A(\mathbf{x}, \mathbf{R}) + V(\mathbf{x}, \mathbf{R})] \phi_i(\mathbf{x}, \mathbf{R}) = E_i(\mathbf{R}) \phi_i(\mathbf{x}, \mathbf{R}) \tag{4.7}$$

which form an orthonormal (orthogonal and normalized) set of functions, such that

$$\langle \phi_j | \phi_i \rangle = \int \phi_j^*(\mathbf{x}, \mathbf{R}) \phi_i(\mathbf{x}, \mathbf{R}) \, d\mathbf{x} = \delta_{ij} \tag{4.8}$$

where δ_{ij} is the Kronecker delta symbol ($\delta_{ij} = 1$ if $i = j$, $\delta_{ij} = 0$ if $i \neq j$). Substituting (4.6) in the Schrödinger equation (4.5) and projecting out ϕ_j by operating with $\int d\mathbf{x}\, \phi_j^*(\mathbf{x}, \mathbf{R})$ on both sides of the equation, we obtain

$$\left[-\frac{\nabla_R^2}{2\mu} + E_j(\mathbf{R}) - E \right] F_j(\mathbf{R})$$
$$= \sum_i \left[\frac{\langle \phi_j | \nabla_\mathbf{R} | \phi_i \rangle \cdot \nabla_\mathbf{R} F_i(\mathbf{R})}{\mu} + \frac{\langle \phi_j | \nabla_R^2 | \phi_i \rangle F_i(\mathbf{R})}{2\mu} \right] \tag{4.9}$$

Were it not for the terms on the right-hand side of equation (4.9), we would have succeeded in separating Schrödinger's equation into (4.7) for the electronic motion at a given value of **R**, and (4.9) for the relative motion on a given *electronic potential energy surface*, $E_j(\mathbf{R})$.

The approximation of neglecting the coupling between the electronic and relative motions, embodied in the terms on the right-hand side of (4.9), is known as the 'adiabatic' or Born–Oppenheimer approximation. These terms give rise to transitions between potential energy curves, are responsible for charge transfer processes (Chapter 8), and can be responsible for fine structure transitions in atoms and ions (Chapter 6), induced in collisions with other atomic or molecular species. In the discussion of rotational excitation processes which follows, the collision will be assumed to take place along a single adiabatic potential energy curve.

52 The rotational excitation of molecules

4.3 The scattering of an atom by a rigid rotator

The theory of the scattering of a structureless particle by a rigid rotator (rotor) was given its first quantum mechanical formulation by Arthurs and Dalgarno [44]. Their approach is applicable to collisions between any particle without internal structure, or whose internal structure may be neglected, and a two-particle system possessing internal angular momentum. The degrees of freedom of such a system may be defined by the three polar coordinates of the atom A in a coordinate system whose origin is at the centre of mass M of the rotor BC and which is fixed in space, together with the two polar angles defining the orientation of BC in this same coordinate frame. The atom may be considered to move with reduced mass $\mu = m_A(m_B + m_C)/(m_A + m_B + m_C)$ relative to the centre of mass of the rotor.

An alternative approach [45] is to define the polar angles of BC relative to a coordinate system in which the Z-axis coincides with MA and rotates in space in the course of the collision. This 'body-fixed' frame of reference is the more natural choice from the viewpoint of the interaction potential, which depends on R and θ' only; see Fig. 4.2. However, as B and C are now moving relative to a coordinate system which is itself rotating, Coriolis forces arise, in addition to centrifugal forces. We shall consider the relative merits of these two coordinate frames in the discussion below.

An interesting analogy may be drawn between atom–rigid rotor scattering, as formulated by Arthurs and Dalgarno [44], and e^-–H scattering, as formulated by Percival and Seaton [46] and considered in Chapter 9. If electron exchange is neglected, the two problems become formally very similar. Indeed, as we shall see below, the algebraic coefficients that arise in the quantum mechanical treatment of atom–rigid rotor scattering are identical to coefficients tabulated by Percival and Seaton.

Let us denote by (Θ, Φ) the orientation of the body-fixed (BF) Z-axis relative to the space-fixed (SF) frame, xyz. The polar coordinates of particle A in the SF frame are then

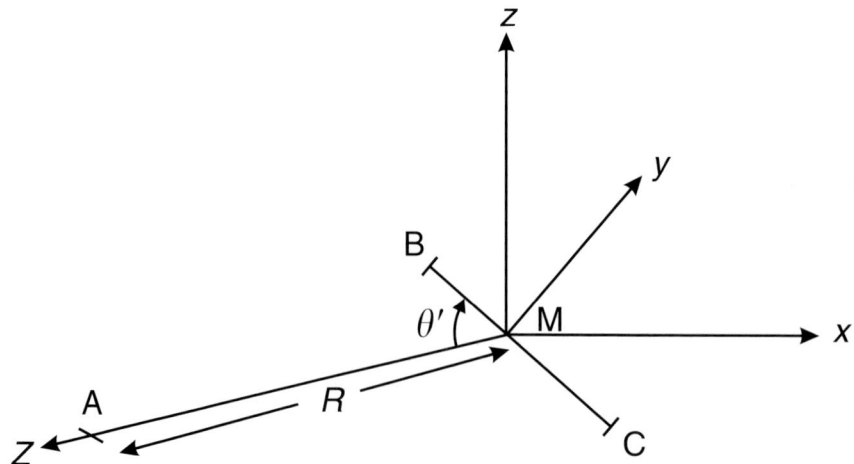

Figure 4.2 Defining the space-fixed coordinate system xyz and the body-fixed Z-axis for the collision between an atom A and a rigid rotor BC, whose centre of mass is M; R and θ' are sometimes called 'Jacobi coordinates'.

4.3 The scattering of an atom by a rigid rotator

(R, Θ, Φ). As noted above, the polar angles of BC may be expressed relative to either the SF or the BF frame. Let us denote these angles (θ, ϕ) and (θ', ϕ'), respectively.

Calculations are facilitated by expressing the wave function of the rotor in terms of a complete set of orthonormal functions of the polar angles (θ, ϕ) or (θ', ϕ'). The normalized spherical harmonics, Y, form such a set of functions. Denoting the angular momentum quantum number of the rotor by j and its projection on the SF z-axis by m, we have that

$$Y_{jm}(\theta, \phi) = (-1)^m \left[\frac{(2j+1)(l-m)!}{4\pi (l+m)!} \right]^{\frac{1}{2}} P_j^m(\cos\theta) e^{im\phi}, \quad (m \geq 0) \tag{4.10}$$

with

$$Y_{j,-m}(\theta, \phi) = (-1)^m Y_{jm}^*(\theta, \phi) \tag{4.11}$$

and where $P_j^m(\cos\theta)$ is an associated Legendre polynomial [47]. In the BF coordinate system, the corresponding set of functions is $Y_{j\Omega}(\theta', \phi')$, where Ω is the projection of $\mathbf{j}$, the angular momentum of the rotor, on the BF Z-axis. In order to establish the relationship between $Y_{jm}(\theta, \phi)$ and $Y_{j\Omega}(\theta', \phi')$, it is necessary to introduce the concepts of the *Euler angles* and the *rotation matrix*, $\mathbf{D}$.

The Euler angles (α, β, γ) define a sequence of three rotations, successively through α about the z-axis, through β about the new y-axis, and through γ about the new z-axis, which takes the SF coordinate system into the BF system. The rotations are taken in the positive sense, that is, in the sense of inserting a right-handed screw. In the example that we are considering, it may be verified that the Euler angles are $\alpha = \Phi$, $\beta = \Theta$, and γ is arbitrary, usually taken to be zero. The elements of the rotation matrix, $\mathbf{D}$, are defined by Rose [48] and Brink and Satchler [49] as

$$D_{m'm}^j(\alpha, \beta, \gamma) = \langle jm' | e^{-i\alpha j_z} e^{-i\beta j_y} e^{-i\gamma j_z} | jm \rangle \tag{4.12}$$

where $|jm\rangle = Y_{jm}(\theta, \phi)$ and j_z, j_y are components of the angular momentum operator in the SF system, where

$$j_x^2 + j_y^2 + j_z^2 = \mathbf{j}^2 \tag{4.13}$$

Using the quantum theory of angular momentum, (4.12) may be written

$$D_{m'm}^j(\alpha, \beta, \gamma) = e^{-im'\alpha} d_{m'm}^j(\beta) e^{-im\gamma} \tag{4.14}$$

where

$$d_{m'm}^j(\beta) = \langle jm' | e^{-i\beta j_y} | jm \rangle \tag{4.15}$$

Explicit expressions for $d_{m'm}^j(\beta)$ are given by Rose [48] and Brink and Satchler [49]. In another standard text on angular momentum theory, Edmonds [50] defines a rotation matrix $\mathcal{D}$ whose elements relate to Rose and Brink and Satchler through

$$\mathcal{D}_{m'm}^j(\alpha, \beta, \gamma) = D_{mm'}^{j*}(\alpha, \beta, \gamma) \tag{4.16}$$

Thus, $\mathcal{D}$ is the transposed complex conjugate of $\mathbf{D}$. Either of these conventions may be adopted but must be adhered to. We shall adopt the definition (4.12) of Rose and Brink and Satchler, in which case

$$Y_{jm}(\theta,\phi) = \sum_{\Omega} D^{j*}_{m\Omega}(\Phi,\Theta,0) Y_{j\Omega}(\theta',\phi') \tag{4.17}$$

and the inverse relation

$$Y_{j\Omega}(\theta',\phi') = \sum_{m} D^{j}_{m\Omega}(\Phi,\Theta,0) Y_{jm}(\theta,\phi) \tag{4.18}$$

apply. The functions (4.17) are eigenfunctions of $\mathbf{j}^2$ and j_z, with eigenvalues $j(j+1)$ and m, respectively, whereas the functions (4.18) are eigenfunctions of $\mathbf{j}^2$ and j_Z, with eigenvalues $j(j+1)$ and Ω.

Conservation laws are at the heart of physics, and it is advantageous, when solving a dynamical problem, to make use of the fact that the total angular momentum is conserved. This symmetry property arises from the invariance of the hamiltonian describing the dynamical system under rotations of the system in space. Put in other words, the orientation of the SF coordinate system may be chosen arbitrarily. In the problem under consideration, the total angular momentum, $\mathbf{J}$, is composed of the angular momentum of the rotor, $\mathbf{j}$, and the angular momentum of the atom relative to the rotor, $\mathbf{l}$:

$$\mathbf{J} = \mathbf{j} + \mathbf{l}$$

$$J_z = j_z + l_z$$

Eigenfunctions of $\mathbf{j}^2$ and j_z have already been given as $Y_{jm}(\theta,\phi)$. Similarly, the eigenfunctions of $\mathbf{l}^2$ and l_z are $Y_{lm_l}(\Theta,\Phi)$, with eigenvalues $l(l+1)$ and m_l, respectively. It follows that the product $Y_{jm}(\theta,\phi)Y_{lm_l}(\Theta,\Phi)$ is an eigenfunction of $\mathbf{j}^2, j_z, \mathbf{l}^2$ and l_z, but it is *not* an eigenfunction of the total angular momentum operators $\mathbf{J}^2$ or J_z. However, such functions are readily formed as

$$\mathcal{Y}_{jlJM}(\theta,\phi;\Theta,\Phi) = \sum_{mm_l} C^{jlJ}_{mm_lM} Y_{jm}(\theta,\phi) Y_{lm_l}(\Theta,\Phi) \tag{4.19}$$

where $M = m + m_l$ and $C^{jlJ}_{mm_lM}$ is a *Clebsch–Gordan coefficient*, which is related to the Wigner 3j-symbol through

$$C^{jlJ}_{mm_lM} = (-1)^{j-l+M} (2J+1)^{\frac{1}{2}} \begin{pmatrix} j & l & J \\ m & m_l & -M \end{pmatrix} \tag{4.20}$$

[51]. Using equation (4.17) and the following relationship between a spherical harmonic and an element of the rotation matrix

$$Y_{lm_l}(\Theta,\Phi) = \left(\frac{2l+1}{4\pi}\right)^{\frac{1}{2}} D^{l*}_{m_l 0}(\Phi,\Theta,0) \tag{4.21}$$

4.3 The scattering of an atom by a rigid rotator

the eigenfunctions (4.19) may be expressed as

$$\mathcal{Y}_{jlJM}(\theta,\phi;\Theta,\Phi) = \left(\frac{2l+1}{4\pi}\right)^{\frac{1}{2}} \sum_{\substack{mm_l \\ \Omega}} C^{jlJ}_{mm_lM} Y_{j\Omega}(\theta',\phi')$$

$$\times D^{j*}_{m\Omega}(\Phi,\Theta,0) D^{l*}_{m_l0}(\Phi,\Theta,0) \quad (4.22)$$

The Clebsch–Gordan series [48] tells us that

$$D^{j}_{m\Omega}(\Phi,\Theta,0) D^{l}_{m_l0}(\Phi,\Theta,0) = \sum_{J'} C^{jlJ'}_{mm_lM} C^{jlJ'}_{\Omega 0\Omega} D^{J'}_{M\Omega}(\Phi,\Theta,0) \quad (4.23)$$

and, as the Clebsch–Gordan coefficients are real, equation (4.22) becomes

$$\mathcal{Y}_{jlJM}(\theta,\phi;\Theta,\Phi) = \left(\frac{2l+1}{4\pi}\right)^{\frac{1}{2}} \sum_{\substack{mm_l \\ J'\Omega}} C^{jlJ}_{mm_lM} C^{jlJ'}_{mm_lM} C^{jlJ'}_{\Omega 0\Omega}$$

$$\times Y_{j\Omega}(\theta',\phi') D^{J'*}_{M\Omega}(\Phi,\Theta,0) \quad (4.24)$$

The Clebsch–Gordan coefficients satisfy orthonormality relations [51], one of which is

$$\sum_{mm_l} C^{jlJ}_{mm_lM} C^{jlJ'}_{mm_lM'} = \delta_{JJ'}\delta_{MM'} \quad (4.25)$$

and so equation (4.24) reduces to

$$\mathcal{Y}_{jlJM}(\theta,\phi;\Theta,\Phi) = \left(\frac{2l+1}{2J+1}\right)^{\frac{1}{2}} \sum_{\Omega} C^{jlJ}_{\Omega 0\Omega} \mathcal{Z}_{j\Omega JM}(\theta',\phi';\Theta,\Phi) \quad (4.26)$$

where

$$\mathcal{Z}_{j\Omega JM}(\theta',\phi';\Theta,\Phi) = \left(\frac{2J+1}{4\pi}\right)^{\frac{1}{2}} D^{J*}_{M\Omega}(\Phi,\Theta,0) Y_{j\Omega}(\theta',\phi') \quad (4.27)$$

is an eigenfunction of $\mathbf{J}^2$ and J_Z. Equation (4.26) specifies the unitary transformation which relates the eigenfunctions of $\mathbf{J}^2$ in the SF frame, $\mathcal{Y}_{jlJM}(\theta,\phi;\Theta,\Phi)$, to the corresponding eigenfunctions in the BF frame, $\mathcal{Z}_{j\Omega JM}(\theta',\phi';\Theta,\Phi)$. An alternative and more compact way of writing equation (4.26) is

$$|jlJM\rangle = |j\Omega JM\rangle\langle j\Omega JM | jlJM\rangle \quad (4.28)$$

with an implied summation over the index Ω, and where

$$\langle j\Omega JM | jlJM\rangle = \left(\frac{2l+1}{2J+1}\right)^{\frac{1}{2}} C^{jlJ}_{\Omega 0\Omega} \quad (4.29)$$

is seen to be independent of the projection quantum number, $M = m + m_l$.

Equation (4.26), or its alternative form, equation (4.28), is an important result, enabling quantities evaluated in the rotating (BF) frame, which is more natural for expressing the interaction between A and BC, to be transformed into the laboratory (SF) frame, in which measurements are made. Note that, as the BF Z-axis is taken to be coincident with MA, the projection of the orbital angular momentum **l** of A relative to BC on this axis is zero, as **l** is perpendicular to MA. It follows that $j_Z = J_Z$. These facts are embodied in the Clebsch–Gordan coefficient, $C^{jlJ}_{\Omega 0 \Omega}$, which appears in equation (4.29): the same projection quantum number, Ω, is associated with both j and J.

Another conservation law which may be used is associated with the parity of the eigenfunctions representing the system A + BC. The corresponding symmetry operation is the inversion of the coordinates of all particles (A, B and C) in the origin of the SF coordinate frame. The hamiltonian is invariant under this operation; the corresponding operator, P, gives rise to the following transformation of SF coordinates:

$$\theta \to \pi - \theta, \phi \to \pi + \phi$$

$$\Theta \to \pi - \Theta, \Phi \to \pi + \Phi$$

Under this same operation, the BF coordinates transform as

$$\theta' \to \theta', \phi' \to \pi - \phi'$$

[52]. Carrying out these transformations of angles, we find that the eigenfunctions (4.19) of $\mathbf{J}^2$ and J_z behave as

$$P\mathcal{Y}_{jlJM}(\theta, \phi; \Theta, \Phi) = \mathcal{Y}_{jlJM}(\pi - \theta, \pi + \phi; \pi - \Theta, \pi + \Phi)$$

$$= (-1)^{j+l} \mathcal{Y}_{jlJM}(\theta, \phi; \Theta, \Phi) \quad (4.30)$$

where use has been made of the properties of the spherical harmonics (4.10). Thus, $\mathcal{Y}_{jlJM}$ is an eigenfunction not only of $\mathbf{J}^2$ and J_z, with eigenvalues $J(J+1)$ and M, respectively, but also of the inversion operator, P, with eigenvalue $(-1)^{j+l}$. This latter eigenvalue is denoted by the parity, $p = \pm 1$, of the wave function.

Regarding the eigenfunctions (4.27) of $\mathbf{J}^2$ and J_Z, the situation is somewhat more complicated. In this case, we find that

$$P\mathcal{Z}_{j\Omega JM}(\theta', \phi'; \Theta, \Phi) = \mathcal{Z}_{j\Omega JM}(\theta', \pi - \phi'; \pi - \Theta, \pi + \Phi)$$

$$= (-1)^J \mathcal{Z}_{j,-\Omega JM}(\theta', \phi'; \Theta, \Phi) \quad (4.31)$$

As Ω appears on the left-hand side, and $-\Omega$ on the right-hand side, of equation (4.31), it follows that, while $\mathcal{Z}_{j\Omega JM}$ is an eigenfunction of $\mathbf{J}^2$ and J_Z, with eigenvalues $J(J+1)$ and Ω, respectively, it is it not an eigenfunction of P. However, such eigenfunctions may readily be formed as the linear combinations

$$\mathcal{Z}_{j\bar{\Omega}\epsilon JM} = \frac{\left(\mathcal{Z}_{j\bar{\Omega}JM} + \epsilon \mathcal{Z}_{j,-\bar{\Omega}JM}\right)}{[2(1+\delta_{\bar{\Omega}0})]^{\frac{1}{2}}} \quad (4.32)$$

4.3 The scattering of an atom by a rigid rotator

where $\bar{\Omega} = |\Omega|$ and $\epsilon = \pm 1$. The factor $\left[2(1+\delta_{\bar{\Omega}0})\right]^{\frac{1}{2}}$ ensures that the eigenfunctions are correctly normalized for all possible values of $\bar{\Omega}$, i.e. for $\bar{\Omega} \geq 0$. It may be seen from equation (4.32) that, when $\Omega = 0$, only $\epsilon = +1$ is allowed: the eigenfunction vanishes when $\epsilon = -1$. The functions $\mathcal{Z}_{j\bar{\Omega}\epsilon JM}$ are eigenfunctions of the inversion operator, P, with eigenvalue $p' = (-1)^J \epsilon$. When $\epsilon = (-1)^{j+l+J}$, that is, when $p' = p$, $\mathcal{Y}_{jlJM}$ and $\mathcal{Z}_{j\bar{\Omega}\epsilon JM}$ are related through

$$\mathcal{Y}_{jlJM}(\theta,\phi;\Theta,\Phi) = \sum_{\bar{\Omega}} \left[\frac{2(2l+1)}{(1+\delta_{\bar{\Omega}0})(2J+1)}\right]^{\frac{1}{2}} C^{jlJ}_{\bar{\Omega}0\bar{\Omega}}$$

$$\times \mathcal{Z}_{j\bar{\Omega}\epsilon JM}(\theta',\phi';\Theta,\Phi) \quad (4.33)$$

In matrix notation, equation (4.33) becomes

$$|jlJM\rangle = |j\bar{\Omega}\epsilon JM\rangle\langle j\bar{\Omega}\epsilon JM|jlJM\rangle \quad (4.34)$$

with $\epsilon = (-1)^{j+l+J}$ and an implied summation over $\bar{\Omega} \geq 0$. It follows from the properties of the Clebsch–Gordan coefficients [51] that, when $\bar{\Omega} = 0$, $C^{jlJ}_{\bar{\Omega}0\bar{\Omega}} = C^{jlJ}_{000} = 0$ unless $j+l+J$ is even; this implies that only $\epsilon = (-1)^{j+l+J} = +1$ is allowed when $\bar{\Omega} = 0$, a condition already noted above.

A rotation of the coordinate system, from the SF to the BF frame, leaves the parity of the wave function unchanged. As a consequence, equation (4.33) [or equivalently (4.34)] is applicable only when $p' = p$. With $p' = p = (-1)^J \epsilon$, equation (4.32) may be written in the form

$$\mathcal{Z}_{j\bar{\Omega}pJM} = \frac{\left(\mathcal{Z}_{j\bar{\Omega}JM} + (-1)^J p \mathcal{Z}_{j,-\bar{\Omega}JM}\right)}{\left[2(1+\delta_{\bar{\Omega}0})\right]^{\frac{1}{2}}}$$

Similarly, as $p = (-1)^{j+l}$, we may define

$$\mathcal{Y}_{jlpJM} \equiv \mathcal{Y}_{jlJM}$$

where the parity subscript is written explicitly. Either the SF functions $\mathcal{Y}_{jlpJM}(\theta,\phi;\Theta,\Phi)$ or the BF functions $\mathcal{Z}_{j\bar{\Omega}pJM}(\theta',\phi';\Theta,\Phi)$ are suitable as a basis in which to expand the total wave function, Ψ,

$$\Psi(\hat{\mathbf{r}},\mathbf{R}) = \sum_{jlpJM} \frac{F(jlpJM|R)}{R} \mathcal{Y}_{jlpJM}(\hat{\mathbf{r}};\hat{\mathbf{R}}) \quad (4.35)$$

or

$$\Psi(\hat{\mathbf{r}}',\mathbf{R}) = \sum_{j\bar{\Omega}pJM} \frac{G(j\bar{\Omega}pJM|R)}{R} \mathcal{Z}_{j\bar{\Omega}pJM}(\hat{\mathbf{r}}';\hat{\mathbf{R}}) \quad (4.36)$$

where $\hat{\mathbf{r}} = (\theta,\phi)$, $\hat{\mathbf{r}}' = (\theta',\phi')$, $\hat{\mathbf{R}} = (\Theta,\Phi)$ and $\mathbf{R} = (R,\Theta,\Phi)$ denote polar coordinates.

The rotational excitation of molecules

The Schrödinger equation (4.5) may be written

$$(H - E)\Psi = 0 \qquad (4.37)$$

where the hamiltonian is given by

$$H = \frac{\mathbf{j}^2}{2I} - \frac{\nabla_R^2}{2\mu} + V(R,\theta') \qquad (4.38)$$

The first term on the right-hand side of equation (4.38) will be recognized as the rotational energy of a rigid rotor. We recall that the rotational angular momentum of a rotor is $j = I\omega$, where I is the moment of inertia and ω is the angular velocity. The associated kinetic energy is $T = I\omega^2/2 = \mathbf{j}^2/(2I)$. The term $V(R,\theta')$ denotes the potential of interaction between A and BC on a given potential energy surface and was written above [cf. equation (4.9)] as $E_j(\mathbf{R})$. The energy of interaction between an atom A and a rigid rotor BC depends on the BF coordinates R and θ' only; V is independent of ϕ' as the potential is invariant under rotations of the internuclear axis BC sbout the BF Z-axis. For the purposes of the subsequent analysis, $V(R,\theta')$ is expanded over a complete set of functions of the angular variable, θ',

$$V(R,\theta') = \sum_{\lambda=0}^{\infty} v_\lambda(R) P_\lambda(\cos\theta') \qquad (4.39)$$

where P_λ is the Legendre polynomial [47]. In practice, the summation in (4.39) is truncated at a finite and sometimes small value of λ.

The second term on the right-hand side of equation (4.38) represents the kinetic energy of the relative motion of A and BC and may be separated into radial and angular parts:

$$-\frac{1}{2\mu}\nabla_R^2 = -\frac{1}{2\mu R}\frac{\partial^2}{\partial R^2}R + \frac{\mathbf{l}^2}{2\mu R^2} \qquad (4.40)$$

We recall that $\mu = m_A(m_B + m_C)/(m_A + m_B + m_C)$ is the reduced mass of the system A + BC. Thus, equation (4.37) becomes

$$\left[\frac{\mathbf{j}^2}{2I} - \frac{1}{2\mu R}\frac{\partial^2}{\partial R^2}R + \frac{\mathbf{l}^2}{2\mu R^2} + V(R,\theta') - E\right]\Psi = 0 \qquad (4.41)$$

with Ψ given by equation (4.35) or equation (4.36). Recalling that

$$\mathbf{j}^2 \mathcal{Y}_{jlpJM} = j(j+1)\mathcal{Y}_{jlpJM}$$

and

$$\mathbf{j}^2 \mathcal{Z}_{j\bar{\Omega}pJM} = j(j+1)\mathcal{Z}_{j\bar{\Omega}pJM}$$

equation (4.41) may be written

$$\left[-\frac{1}{R}\frac{\partial^2}{\partial R^2}R + \frac{\mathbf{l}^2}{R^2} + 2\mu V(R,\theta') - k_j^2\right]\Psi = 0 \qquad (4.42)$$

4.3 The scattering of an atom by a rigid rotator

where

$$k_j^2 = 2\mu[E - Bj(j+1)] \tag{4.43}$$

and $B = 1/(2I)$ is the rotational constant of the molecule BC.

We may now make use of the orthonormality properties of the basis functions, $\mathcal{Y}_{jlpJM}$ and $\mathcal{Z}_{j\bar{\Omega}pJM}$:

$$\int \mathcal{Y}_{jlpJM}^*(\hat{\mathbf{r}};\hat{\mathbf{R}})\mathcal{Y}_{j'l'p'J'M'}(\hat{\mathbf{r}};\hat{\mathbf{R}})\,\mathrm{d}\hat{\mathbf{r}}\,\mathrm{d}\hat{\mathbf{R}} = \delta_{jj'}\delta_{ll'}\delta_{pp'}\delta_{JJ'}\delta_{MM'}$$

and

$$\int \mathcal{Z}_{j\bar{\Omega}pJM}^*(\hat{\mathbf{r}}';\hat{\mathbf{R}})\mathcal{Z}_{j'\bar{\Omega}'p'J'M'}(\hat{\mathbf{r}}';\hat{\mathbf{R}})\,\mathrm{d}\hat{\mathbf{r}}'\,\mathrm{d}\hat{\mathbf{R}} = \delta_{jj'}\delta_{\bar{\Omega}\bar{\Omega}'}\delta_{pp'}\delta_{JJ'}\delta_{MM'}$$

We operate on equation (4.42) from the left with

$$\int \mathrm{d}\hat{\mathbf{r}}\,\mathrm{d}\hat{\mathbf{R}}\,\mathcal{Y}_{jlpJM}^*(\hat{\mathbf{r}};\hat{\mathbf{R}})$$

or

$$\int \mathrm{d}\hat{\mathbf{r}}'\,\mathrm{d}\hat{\mathbf{R}}\,\mathcal{Z}_{j\bar{\Omega}pJM}^*(\hat{\mathbf{r}}';\hat{\mathbf{R}})$$

according to whether the SF or the BF expansion, equation (4.35) or (4.36), is used for the wave function, Ψ. Equation (4.42) then reduces to

$$\left[\frac{\mathrm{d}^2}{\mathrm{d}R^2} - \frac{l(l+1)}{R^2} + k_j^2\right]F(jlpJM|R)$$
$$= 2\mu \sum_{j'l'p'J'M'} \langle jlpJM|V(R,\theta')|j'l'p'J'M'\rangle F(j'l'p'J'M'|R) \tag{4.44}$$

or

$$\left[\frac{\mathrm{d}^2}{\mathrm{d}R^2} + k_j^2\right]G(j\bar{\Omega}pJM|R)$$
$$= 2\mu \sum_{j'\bar{\Omega}'p'J'M'} \langle j\bar{\Omega}pJM|V(R,\theta')$$
$$+ \frac{\mathbf{l}^2}{2\mu R^2}|j'\bar{\Omega}'p'J'M'\rangle G(j'\bar{\Omega}'p'J'M'|R) \tag{4.45}$$

depending on whether the SF or the BF basis functions are used. When deriving equation (4.44), we have made use of the fact that

$$\mathbf{l}^2 \mathcal{Y}_{jlpJM} = l(l+1)\mathcal{Y}_{jlpJM}$$

which gives rise to the centrifugal term, $l(l+1)/R^2$. We use the more compact bra-ket notation on the right-hand sides of equations (4.44) and (4.45).

Equations (4.44) and 4.45) are equivalent, and identical cross-sections should be obtained when these equations are solved without further approximation. Both (4.44) and (4.45) represent sets of ordinary differential equations, which are linear in the functions of the radial coordinate, $F(R)$ or $G(R)$, involve second-order derivatives with respect to R, and are 'coupled' through the matrix elements on the right-hand sides. Powerful numerical techniques have been developed for solving such systems of equations and incorporated in the MOLSCAT [53], HIBRIDON [54] and MOLCOL [55] computer codes.

The use of either equation (4.44) or equation (4.45) has its advantages and drawbacks. In (4.44), the centrifugal term takes a simple form because the SF basis functions are eigenfunctions of the operator $\mathbf{l}^2$. However, in the matrix elements on the right-hand side, the basis functions depend on the SF coordinate, θ, whereas the potential V depends on the BF coordinate, θ'. The evaluation of these matrix elements will be considered below. In (4.45), on the other hand, the matrix elements involving V may be evaluated directly, as the basis functions also depend on BF coordinates. In this case, it is the operator representing the centrifugal potential, $\mathbf{l}^2/(2\mu R^2)$ that poses problems: because the BF coordinate system is itself rotating, not only centrifugal but also Coriolis terms arise when evaluating this operator, the expression for which will be given below.

4.3.1 The space-fixed (SF) basis functions

We consider the matrix elements of the potential,

$$\langle jlpJM|V(R,\theta')|j'l'p'J'M'\rangle$$

$$= \int \mathcal{Y}^*_{jlpJM}(\hat{\mathbf{r}};\hat{\mathbf{R}})V(R,\theta')\mathcal{Y}_{j'l'p'J'M'}(\hat{\mathbf{r}};\hat{\mathbf{R}})\,d\hat{\mathbf{r}}\,d\hat{\mathbf{R}}$$

$$= \sum_\lambda v_\lambda(R) \int \mathcal{Y}^*_{jlpJM}(\hat{\mathbf{r}};\hat{\mathbf{R}})P_\lambda(\cos\theta')\mathcal{Y}_{j'l'p'J'M'}(\hat{\mathbf{r}};\hat{\mathbf{R}})\,d\hat{\mathbf{r}}\,d\hat{\mathbf{R}} \qquad (4.46)$$

where we introduce the expansion (4.39) of the potential V in terms of the Legendre polynomials, P_λ. The integral (4.46) may be evaluated by means of the spherical harmonic addition theorem [48], which states that

$$P_\lambda(\cos\theta') = \frac{4\pi}{2\lambda+1} \sum_{\nu=-\lambda}^{\lambda} Y_{\lambda\nu}(\hat{\mathbf{r}})Y^*_{\lambda\nu}(\hat{\mathbf{R}}) \qquad (4.47)$$

where $Y_{\lambda\nu}$ is a spherical harmonic function, given by (4.10) above. This theorem converts the dependence of the potential on θ', which is the angle between the intramolecular vector $\hat{\mathbf{r}}$ and the intermolecular vector $\hat{\mathbf{R}}$ (see Fig. 4.2), into its dependence on the SF angles $\hat{\mathbf{r}} = (\theta,\phi)$ and $\hat{\mathbf{R}} = (\Theta,\Phi)$ and enables the integrals in equation (4.46) to be carried out. Using the definition (4.19) of the SF basis functions and the composition relations for spherical harmonics [48],

$$\int Y^*_{jm}(\hat{\mathbf{r}})Y_{\lambda\nu}(\hat{\mathbf{r}})Y_{j'm'}(\hat{\mathbf{r}})\,d\hat{\mathbf{r}} = \left[\frac{(2j'+1)(2\lambda+1)}{4\pi(2j+1)}\right]^{\frac{1}{2}} C^{j'\lambda j}_{m'\nu m} C^{j'\lambda j}_{000} \qquad (4.48)$$

4.3 The scattering of an atom by a rigid rotator

and

$$\int Y^*_{lm_l}(\hat{\mathbf{R}}) Y^*_{\lambda\nu}(\hat{\mathbf{R}}) Y_{l'm_{l'}}(\hat{\mathbf{R}}) \, d\hat{\mathbf{R}} = \left[\frac{(2l+1)(2\lambda+1)}{4\pi(2l'+1)}\right]^{\frac{1}{2}} C^{l\lambda l'}_{m_l \nu m_{l'}} C^{l\lambda l'}_{000} \quad (4.49)$$

Equation (4.46) becomes

$$\langle jlpJM | V(R,\theta') | j'l'p'J'M' \rangle$$

$$= \sum_{\lambda\nu} v_\lambda(R) \left[\frac{(2j'+1)(2l+1)}{(2j+1)(2l'+1)}\right]^{\frac{1}{2}} C^{j'\lambda j}_{000} C^{l\lambda l'}_{000}$$

$$\times \sum_{mm_l m' m_{l'}} C^{jlJ}_{mm_l M} C^{j'\lambda j}_{m'\nu m} C^{j'l'J'}_{m'm_{l'}M'} C^{l\lambda l'}_{m_l \nu m_{l'}} \quad (4.50)$$

Using angular momentum recoupling theory (see, for example, [51]), equation (4.50) may be expressed in terms of a Racah coefficient, W:

$$\langle jlpJM | V(R,\theta') | j'l'p'J'M' \rangle$$

$$= \delta_{JJ'}\delta_{MM'}(-1)^{j+j'-J} \sum_\lambda v_\lambda(R) \frac{[(2j+1)(2l+1)(2j'+1)(2l'+1)]^{\frac{1}{2}}}{(2\lambda+1)}$$

$$\times C^{jj'\lambda}_{000} C^{ll'\lambda}_{000} W(jlj'l'; J\lambda) \quad (4.51)$$

The Racah coefficient is related to the 6j-symbol of Wigner through

$$W(jlj'l'; J\lambda) = (-1)^{j+l+j'+l'} \begin{Bmatrix} j & l & J \\ l' & j' & \lambda \end{Bmatrix} \quad (4.52)$$

and is an algebraic quantity that is readily evaluated for given values of the arguments. The Kronecker δ symbols appearing in (4.51) ensure the conservation of the total angular momentum, J, and its projection M, on the SF z-axis. As $C^{jj'\lambda}_{000}$ and $C^{ll'\lambda}_{000}$ vanish identically unless $j+j'+\lambda$ and $l+l'+\lambda$, respectively, are even, we see that

$$(-1)^{j+j'+\lambda+l+l'+\lambda} = +1 = (-1)^{j+l+j'+l'} \quad (4.53)$$

because λ is an integer. It follows that

$$p = (-1)^{j+l} = (-1)^{j'+l'} = p' \quad (4.54)$$

i.e. the parity is conserved. Using these conservation relations, we may write equation (4.51) in the more compact form

$$\langle jlpJM | V(R,\theta') | j'l'pJM \rangle = \sum_\lambda v_\lambda(R) f_\lambda(jl, j'l'; J) \quad (4.55)$$

where the algebraic coefficient

$$f_\lambda(jl,j'l';J) = (-1)^{j+j'-J} \frac{[(2j+1)(2l+1)(2j'+1)(2l'+1)]^{\frac{1}{2}}}{(2\lambda+1)}$$

$$\times C^{jj'\lambda}_{000} C^{ll'\lambda}_{000} W(jlj'l';J\lambda) \quad (4.56)$$

is independent of the projection quantum number M. The coefficients $f_\lambda(jl,j'l';J)$ were first introduced by Percival and Seaton [46], who were concerned with e$^-$–H scattering (see Chapter 9), and are often referred to as 'Percival–Seaton coefficients'. The coupled equations (4.44) may now be written

$$\left[\frac{d^2}{dR^2} - \frac{l(l+1)}{R^2} + k_j^2\right] F(jlpJ|R)$$

$$= 2\mu \sum_{j'l'\lambda} v_\lambda(R) f_\lambda(jl,j'l';J) F(j'l'pJ|R) \quad (4.57)$$

where the index M has been dropped, as the equations are independent of this quantum number.

It is instructive to consider the form of these equations for $\lambda = 0$. In this case, the Clebsch–Gordan coefficients in (4.56) are non-vanishing only when $j = j'$ and $l = l'$, and hence no collisional coupling between different rotational states of the molecule BC can occur. The term with $\lambda = 0$ in the interaction potential (4.39) is *angle independent*, as $P_0(\cos\theta') = 1$, and cannot induce rotational excitation (or de-excitation) of the molecule; $v_0(R)$ contributes only to *elastic scattering* of A on BC. Terms in the potential with $\lambda \geq 1$, on the other hand, can give rise to rotational transitions in the molecule, subject to the *triangular inequalities* $|j - j'| \leq \lambda \leq j + j'$ and the requirement that $j + j' + \lambda$ should be an even integer. Thus, if $j = 0$, $\lambda = j' = j' - j \equiv \Delta j$. In the CO molecule, for example, excitation from the ground state, $j = 0$, to $j' = 1$ is induced by the term with $\lambda = 1$ in the interaction potential. Similarly, direct excitation from $j = 0$ to $j' = 2$ is induced by the following term, with $\lambda = 2$, and so on. The absolute magnitudes of successive coefficients $v_\lambda(R)$ in the expansion of the potential (4.39) tend to decrease as λ increases, and the probability of transitions involving increasing values of Δj becomes progressively smaller.

In the case of *homonuclear* molecules, where B and C are identical (H_2, N_2, O_2, ...), the interaction potential is clearly invariant under exchange of B and C, equivalent to the operation $\theta' \to \pi - \theta'$. As $\cos(\pi - \theta') = -\cos\theta'$, and

$$P_\lambda(-\cos\theta') = P_\lambda(\cos\theta')$$

when λ is even, and

$$P_\lambda(-\cos\theta') = -P_\lambda(\cos\theta')$$

when λ is odd, it follows that only those terms with even values of λ appear in the interaction potential. As $j + j' + \lambda$ must also be an even integer, collisional transitions between even and odd values of the rotational quantum number j cannot occur.

4.3.2 The body-fixed (BF) basis functions

We must evaluate the matrix elements of the *effective potential*, $V_{\text{eff}}(r, \theta')$, comprising the interaction potential, $V(R, \theta')$, and the centrifugal potential, $\mathbf{l}^2/(2\mu R^2)$:

$$\langle j\bar{\Omega}pJM | V(R, \theta') + \frac{\mathbf{l}^2}{(2\mu R^2)} | j'\bar{\Omega}'p'J'M'\rangle$$

$$= \int \mathcal{Z}^*_{j\bar{\Omega}pJM}(\hat{\mathbf{r}}'; \hat{\mathbf{R}}) \left[V(R, \theta') + \frac{\mathbf{l}^2}{(2\mu R^2)} \right] \mathcal{Z}_{j'\bar{\Omega}'p'J'M'}(\hat{\mathbf{r}}'; \hat{\mathbf{R}}) \, \mathrm{d}\hat{\mathbf{r}}' \, \mathrm{d}\hat{\mathbf{R}} \quad (4.58)$$

The contribution of the interaction potential to this integral is readily evaluated. Using the definition of the BF basis functions (4.27), the relation

$$P_\lambda(\cos\theta') = \left(\frac{4\pi}{2\lambda + 1}\right)^{\frac{1}{2}} Y_{\lambda 0}(\theta', \phi') \quad (4.59)$$

the orthogonality relation for the rotation matrix elements

$$\int D^J_{M\Omega}(\Phi, \Theta, 0) D^{J'*}_{M'\Omega}(\Phi, \Theta, 0) \sin\Theta \, \mathrm{d}\Theta \, \mathrm{d}\Phi = \delta_{JJ'}\delta_{MM'}\left(\frac{4\pi}{2J+1}\right) \quad (4.60)$$

and the composition relation for spherical harmonics

$$\int Y^*_{j\Omega}(\hat{\mathbf{r}}') Y_{\lambda 0}(\hat{\mathbf{r}}') Y_{j'\Omega'}(\hat{\mathbf{r}}') \, \mathrm{d}\hat{\mathbf{r}}' = \delta_{\Omega\Omega'}\left[\frac{(2j'+1)(2\lambda+1)}{4\pi(2j+1)}\right]^{\frac{1}{2}} C^{j'\lambda j}_{\Omega 0 \Omega} C^{j'\lambda j}_{000} \quad (4.61)$$

we obtain

$$\langle j\bar{\Omega}pJM | V(R, \theta') | j'\bar{\Omega}'p'J'M'\rangle$$

$$= \delta_{\bar{\Omega}\bar{\Omega}'}\delta_{pp'}\delta_{JJ'}\delta_{MM'}(-1)^{\bar{\Omega}} \sum_\lambda v_\lambda(R) \frac{[(2j+1)(2j'+1)]^{\frac{1}{2}}}{(2\lambda+1)}$$

$$\times C^{jj'\lambda}_{000} C^{jj'\lambda}_{\bar{\Omega}, -\bar{\Omega} 0} \quad (4.62)$$

Equation (4.62) incorporates the same conservation properties ($J = J'$, $M = M'$, $p = p'$) as those encountered above in the discussion of the matrix elements of the potential in the SF representation. In addition, we have the relation $\bar{\Omega} = \bar{\Omega}'$. This latter property arises from the invariance of the potential under rotations about the BF Z-axis, that is, from the fact that V is independent of ϕ'. The torque about the Z-axis is $\Gamma_Z = -\partial V/\partial \phi'$, and it follows that the component of the angular momentum, Ω, about this same axis cannot be modified by the interaction potential. On the other hand, the angular momentum operator, $\mathbf{l}^2$, *can* change the value of Ω. Using the quantum theory of angular momentum, it may be shown that the non-vanishing matrix elements of the centrifugal potential operator in equation (4.58) are given by

$$\langle j\bar{\Omega}pJM | \frac{\mathbf{l}^2}{2\mu R^2} | j\bar{\Omega}pJM \rangle$$

$$= \frac{J(J+1) + j(j+1) - 2\bar{\Omega}^2}{2\mu R^2} \quad (4.63)$$

and

$$\langle j\bar{\Omega}pJM | \frac{\mathbf{l}^2}{2\mu R^2} | j, \bar{\Omega} \pm 1, pJM \rangle$$

$$= -(1 + \delta_{\bar{\Omega}0})^{\frac{1}{2}} (1 + \delta_{\bar{\Omega}\pm 1, 0})^{\frac{1}{2}}$$

$$\times \frac{[J(J+1) - \bar{\Omega}(\bar{\Omega} \pm 1)]^{\frac{1}{2}} [j(j+1) - \bar{\Omega}(\bar{\Omega} \pm 1)]^{\frac{1}{2}}}{2\mu R^2} \quad (4.64)$$

[45, 52, 56]. These results are to be compared with $l(l+1)/(2\mu R^2)$, the corresponding matrix elements when SF functions are used. The reason for the additional complexity of equations (4.63) and (4.64) – Coriolis forces in the BF frame – has already been mentioned.

Inspection of equations (4.62–4.64) shows the matrix elements of both the interaction potential and the centrifugal potential to be independent of the projection quantum number, M. Accordingly, the coupled equations (4.45) may be written

$$\left[\frac{d^2}{dR^2} + k_j^2 \right] G(j\bar{\Omega}pJ|R)$$

$$= 2\mu \sum_{j'\bar{\Omega}'} V_{\text{eff}}(j\bar{\Omega}, j'\bar{\Omega}'; J|R) G(j'\bar{\Omega}'pJ|R) \quad (4.65)$$

where $V_{\text{eff}}(j\bar{\Omega}, j'\bar{\Omega}'; J|R)$ denotes a matrix element (4.58) of the effective potential,

$$V_{\text{eff}}(R, \theta') = V(R, \theta') + \mathbf{l}^2/(2\mu R^2) \quad (4.66)$$

These equations have a form that is clearly similar to the SF-coupled equations (4.57), and they can be solved by means of the same algorithms.

In summary, Schrödinger's equation for the collision between an atom A and a diatomic molecule BC may be reduced to a set of coupled differential equations. These equations have a similar structure when written in terms of SF or BF coordinates; the Z-axis of the BF coordinate system is chosen to coincide with the vector from the centre of mass M of the molecule to the atom A. The matrix elements of the interaction potential are more tricky to evaluate in the SF frame than in the BF frame, whereas, for the matrix elements of the centrifugal potential, the opposite is true.

One reason for deriving both forms, (4.57) and (4.65), of the coupled equations is that they lend themselves to different types of approximation. The centrifugal potential, $\mathbf{l}^2/(2\mu R^2)$, varies as R^{-2}, whereas, in the collision between a neutral atom and a neutral molecule, the leading term in the potential energy expansion (4.39) varies as R^{-6} at long range.

4.3 The scattering of an atom by a rigid rotator

This comparison suggests that collisions at long range, that is, collisions at large values of the impact parameter and hence of the relative angular momentum, l, might be solved by means of the SF equations with an approximate form of the interaction potential. A possible approximation consists of truncating the potential energy expansion to just a few terms. On the other hand, short-range collisions and small values of l could be solved by means of the BF equations and an approximate form of the centrifugal potential.

Rotationally inelastic collisions involving neutral particles tend to be induced by the interaction potential at short range and lend themselves to approximations based upon the BF-coupled equations; the related approximations will now be presented. Rotationally inelastic collisions form an important category of astrophysical processes. Approximate methods are, and are likely to continue to be, essential aids to solving certain types of molecular collision problems; they are helpful also in understanding the physics involved in such processes.

4.3.3 The coupled states (CS) approximation

The coupled states, or centrifugal decoupling, approximation was introduced by McGuire and Kouri [57]; it has proved to be one of the most successful approximations and has been used extensively in studies of rotational and also vibrational excitation processes. McGuire and Kouri used the BF formulation of the scattering of an atom on a rigid rotor, with the matrix elements of the centrifugal potential, (4.63) and (4.64), approximated by their SF equivalent forms, that is,

$$\langle j\bar{\Omega} pJM | \mathbf{l}^2/(2\mu R^2) | j\bar{\Omega}' pJM \rangle \approx \delta_{\bar{\Omega}\bar{\Omega}'} l(l+1)/(2\mu R^2) \tag{4.67}$$

We see from (4.67) that, subject to this approximation, the centrifugal potential conserves the value of the projection quantum number, $\bar{\Omega}$. As the interaction potential also conserves $\bar{\Omega}$ [cf. equation (4.62)], McGuire and Kouri called this approximation the 'j_z-conserving coupled states approximation'. The BF equations (4.65) are now coupled only through the rotational quantum number, j. Thus, the problem reduces to solving, for each value of $\bar{\Omega}$, a set of differential equations coupled in j, rather than a single set of equations, coupled in both j and $\bar{\Omega}$.

The consequent saving in computing time can be substantial, sometimes rendering feasible calculations which would not be practical otherwise. We recall that, as Ω is the projection of both $\mathbf{j}$ and $\mathbf{J}$ on the BF Z-axis, and $\bar{\Omega} = |\Omega|$, the quantum theory of angular momentum requires that

$$\bar{\Omega} = 0, 1, \ldots, \min(j, J) \quad [p = (-1)^J]$$

$$\bar{\Omega} = 1, 2, \ldots, \min(j, J) \quad [p = -(-1)^J]$$

where $\min(j, J)$ denotes the lesser of j and J. For a problem involving the rotational states $j = 0, 1, \ldots, j_{max}$ and for $J > j_{max}$, the corresponding numbers of coupled equations are

$$N = \sum_{j=0}^{j_{max}} (j+1) = (j_{max}+1)(j_{max}+2)/2 \quad [p = (-1)^J]$$

$$N = \sum_{j=0}^{j_{max}} j = j_{max}(j_{max}+1)/2 \quad [p = -(-1)^J]$$

Depending on the algorithm that is used, the computer time requirement, T, can increase with N as rapidly as $T \propto N^3$. The numerical solution of the complete sets of coupled equations then requires a time $T \propto j_{max}^6/4$ for large j_{max}. On the other hand, when the CS approximation is employed, the number of coupled channels is $(j_{max} + 1)$ for $\bar{\Omega} = 0$, j_{max} for $\bar{\Omega} = 1$, $(j_{max} - 1)$ for $\bar{\Omega} = 2$, and so on, up to 1 channel for $\bar{\Omega} = j_{max}$. The corresponding computer time requirement is

$$T \propto \sum_{j=0}^{j_{max}} [(j+1)^3 + j^3] \sim j_{max}^4/2$$

Thus, the CS approximation is more rapid by a factor $\sim j_{max}^2/2$. Even when modest numbers of rotational levels are involved, $j_{max} \approx 5$, say, using the CS approximation can be about an order of magnitude faster than solving the complete sets of coupled equations.

4.3.4 The infinite order sudden (IOS) approximation

As was recognized by McGuire and Kouri [57], the essence of the CS approximation is to neglect the rotation in space of the BF coordinate system when evaluating the centrifugal potential. If the *sudden* approximation to the rotation of the diatomic molecule is also applicable, that is, if the molecule does not rotate appreciably in the course of the collision, then $\theta' \approx$ constant. This additional approximation is most appropriate for heavy molecules that rotate only slowly and have small rotational constants, B, at collision energies, E, that are large compared with the rotational excitation energies, $Bj(j+1)$, of the levels in question. Then, it follows from (4.43) that

$$k_j^2 = 2\mu[E - Bj(j+1)] \approx 2\mu E \equiv k^2 \quad (4.68)$$

and the scattering equations (4.42) reduce to

$$\left[\frac{d^2}{dR^2} - \frac{l(l+1)}{R^2} - 2\mu V(R, \theta') + k^2\right] G_l(R, \theta') = 0 \quad (4.69)$$

which is to be solved for given values of the parameters θ' and l.

The combination of the centrifugal decoupling and the energy sudden approximations, which leads to equations of the relatively simple form (4.69), is known as the infinite order sudden (IOS) approximation; it derives from the work of Tsien and Pack [58–60] and Pack [61]. This approximation was used extensively in studies of rotational and rovibrational excitation, in a form due to Secrest [62]. The problem of rovibrational excitation will be considered in Chapter 5. The use of the IOS approximation is sometimes necessary, but the CS approximation is certainly to be preferred, whenever its use is feasible.

4.3.5 Boundary conditions

The differential equations derived above, whether exact or approximate, are to be solved subject to appropriate boundary conditions. In order to illustrate the principles involved, we shall consider the form of the boundary conditions which are appropriate when the simplest approximation to the coupled equations, the IOS approximation, is employed.

We are concerned with problems such that

$$V(R, \theta') >> E$$

4.3 The scattering of an atom by a rigid rotator

as $R \to 0$, and

$$V(R, \theta') \sim R^{-n}$$

as $R \to \infty$, where $n \geq 2$ is an integer. The scattering boundary conditions appropriate to this form of potential are

$$G_l(R, \theta') \to 0$$

as $R \to 0$, and

$$G_l(R, \theta') \sim k^{\frac{1}{2}} R[j_l(kR) A_l(\theta') - n_l(kR) B_l(\theta')] \quad (4.70)$$

as $R \to \infty$; j_l and n_l are *spherical Bessel functions* of the first and second kinds, respectively [47]. The coefficients $A_l(\theta')$ and $B_l(\theta')$ are determined by solving numerically the differential equations (4.69), using one of a number of possible algorithms, and fitting to the form (4.70) in the asymptotic region, where the interaction potential has become vanishingly small (in practice, small compared with the collision energy, E).

All relevant information on the scattering process is contained in the quantity

$$S_l(\theta') = 1 + 2\mathrm{i} K_l(\theta')[1 - \mathrm{i} K_l(\theta')]^{-1} \quad (4.71)$$

where $\mathrm{i} = (-1)^{\frac{1}{2}}$ and $K_l(\theta')$ is given by

$$K_l(\theta') = B_l(\theta') A_l^{-1}(\theta') \quad (4.72)$$

Equations (4.70–4.72) are readily generalized to coupled channels scattering, when **K** and **S** are known as the *reactance* and *scattering* matrices, respectively [63].

Equation (4.71), which derives from the sudden approximation to the scattering process, does not in itself yield information on rotationally *inelastic* scattering, that is, on scattering processes involving a change in the rotational state of the molecule. However, this information may be obtained from

$$\begin{aligned} S_l(j\Omega, j'\Omega') &\equiv \langle j\Omega | S_l(\theta') | j'\Omega' \rangle \\ &= \int_{\theta'=0}^{\pi} \int_{\phi'=0}^{2\pi} Y_{j\Omega}^*(\theta', \phi') S_l(\theta') Y_{j'\Omega'}(\theta', \phi') \sin\theta' \, \mathrm{d}\theta' \, \mathrm{d}\phi' \\ &= \delta_{\Omega\Omega'} 2\pi \int_{\theta'=0}^{\pi} Y_{j\Omega}^*(\theta', 0) S_l(\theta') Y_{j'\Omega}(\theta', 0) \sin\theta' \, \mathrm{d}\theta' \end{aligned} \quad (4.73)$$

Such integrals can be evaluated by means of numerical quadrature, having determined $S_l(\theta')$ at the appropriate values of θ' by solving the scattering equation (4.69). Partial cross-sections (i.e. the contributions to the total cross-sections from each value of l) may then be derived from

$$\sigma_l(j \leftarrow j') = \frac{\pi}{k_{j'}^2(2j'+1)} \sum_{\Omega} (2l+1) |S_l(j\Omega, j'\Omega)|^2 \quad (j \neq j') \quad (4.74)$$

and total cross-sections by summing over l.

An alternative form of (4.74), which is better adapted to discussion, is

$$\sigma_l(j \leftarrow j') = \frac{\pi}{k_{j'}^2}(2l+1)P_l(j \leftarrow j') \qquad (4.75)$$

where

$$P_l(j \leftarrow j') = \frac{1}{(2j'+1)}\sum_{\Omega}|S_l(j\Omega, j'\Omega)|^2 \quad (j \neq j') \qquad (4.76)$$

is the *probability* of the $j \leftarrow j'$ rotationally inelastic transition. The *total* cross-section is

$$\sigma(j \leftarrow j') = \sum_l \sigma_l(j \leftarrow j')$$

$$= \frac{\pi}{k_{j'}^2}\sum_l(2l+1)P_l(j \leftarrow j') \qquad (4.77)$$

Equation (4.77) is the quantum mechanical equivalent of the semi-classical expression for a cross-section as an integral of the corresponding transition probability over the impact parameter

$$\sigma(j \leftarrow j') = 2\pi\int_0^\infty P_b(j \leftarrow j')b\,\mathrm{d}b \qquad (4.78)$$

The impact parameter, b, is defined as the distance of closest approach of the atom A to the centre of mass M of the molecule BC, if the atom were to follow a straight-line trajectory. The *semi-classical* transition probability, $P_b(j \leftarrow j')$, is a function of the *classical* impact parameter for a given transition between the *quantized* states of the molecule. To derive (4.77) from (4.78), we note the correspondence between the classical and quantal expressions for the square of the relative angular momentum,

$$2\mu E b^2 = l(l+1) = k^2 b^2$$

where k is the wave number. Differentiating for a given (constant) value of E, we obtain

$$2k^2 b\,\mathrm{d}b = (2l+1)\,\mathrm{d}l$$

where $\mathrm{d}l = 1$ in the quantal limit.

It may be seen from (4.77) or (4.78) that the basic task of either a quantum mechanical or a semi-classical calculation of a collision process is to evaluate the transition probabilities, P_l or P_b, respectively. In subsequent applications of the results, the quantity that is required is the rate coefficient, which is related to the cross-section through

$$\langle \sigma v \rangle_{j \leftarrow j'} = \int_0^\infty v_{j'}\sigma_{j \leftarrow j'}(v_{j'})f(v_{j'},T)\,\mathrm{d}v_{j'} \qquad (4.79)$$

where $v_{j'}$ denotes the relative collision velocity of the atom and molecule in the initial channel, j', and $f(v,T)$ is the Maxwellian velocity distribution at kinetic temperature, T. A Maxwellian distribution is almost always adopted, on the grounds that the timescale for elastic collisions with the most abundant species (H, He or H_2), which tend to thermalize the velocity distribution, is less than the timescale for inelastic collisions, which have smaller cross-sections. Furthermore, in the astrophysical context, the actual velocity distribution cannot be determined, in general. From detailed balance, we have that

$$\sigma(j \leftarrow j')k_{j'}^2 \omega_{j'} = \sigma(j' \leftarrow j)k_j^2 \omega_j \tag{4.80}$$

where the wave number, k_j, is given by

$$k_j = \frac{\mu v_j}{\hbar}$$

and where μ is the reduced mass of the atom–molecule system. The statistical weight (degeneracy), ω_j, is

$$\omega_j = 2j + 1$$

Equation (4.79) may be written as

$$\langle \sigma v \rangle_{j \leftarrow j'} = \frac{\hbar^2}{\omega_{j'}} \left(\frac{2\pi}{\mu^3 k_B T} \right)^{\frac{1}{2}} \int_0^\infty \Omega_{j,j'}(x_{j'}) e^{-x_{j'}} \, dx_{j'} \tag{4.81}$$

where $\hbar = h/(2\pi)$ and h is Planck's constant, k_B is Boltzmann's constant, and

$$\pi \Omega_{j,j'} = \sigma(j \leftarrow j')k_{j'}^2 \omega_{j'}$$

or, from (4.77),

$$\Omega_{j,j'} = (2j' + 1) \sum_l (2l + 1) P_l(j \leftarrow j') \tag{4.82}$$

The dimensionless quantity, $\Omega_{j,j'}$, termed the *collision strength* in the theory of electron–atom scattering, is symmetric in j and j'. Substituting numerical values for the constants in (4.81), we obtain, in $cm^3 s^{-1}$, the units customarily used for rate coefficients,

$$\langle \sigma v \rangle_{j \leftarrow j'} = \frac{1.11 \times 10^{-10}}{\omega_{j'} T^{\frac{1}{2}}} \int_0^\infty \Omega_{j,j'}(x_{j'}) e^{-x_{j'}} \, dx_{j'} \tag{4.83}$$

In problems involving rotational excitation, the numerical value of the summation in (4.82) is typically of the order of 1, and so the corresponding rate coefficients (4.83) are of the order of 10^{-11} $cm^3 s^{-1}$ for $T \approx 100$ K.

4.4 The rotational excitation of non-linear molecules

Many important interstellar molecules have non-linear structures. Fortunately, they generally retain some symmetry properties that can be exploited to make numerical calculations more tractable; these same properties also lead to *collisional propensity rules*. Examples of such molecules are ammonia (NH_3), a symmetric top, water (H_2O), an asymmetric top, and methanol (CH_3OH), which is a near-symmetric top that exhibits internal torsional motion of the CH_3 relative to the OH group. Each of these examples of classes of molecules will now be discussed.

70   *The rotational excitation of molecules*

4.4.1 Symmetric tops

The structure of symmetric top molecules such as ammonia was considered in the classic text of Townes and Shawlow [64]. In the case of NH$_3$, the three hydrogen nuclei form an equilateral triangle, and the nitrogen nucleus is on the line perpendicular to this plane and passing through the geometrical centre of the triangle. The molecule can perform end-over-end rotational motion, which we shall treat using the 'rigid rotor' approximation. We denote the rotational quantum number by j, the projection of the rotational angular momentum on the symmetry axis of the molecule by k, and its projection on the space-fixed z-axis by m. The internal hamiltonian is three-fold symmetric about the symmetry axis of the molecule, i.e. the hamiltonian is invariant under a rotation through 120° about this axis, which is the body-fixed Z-axis.

The three hydrogen nuclei (protons) are identical fermions, which obey Fermi–Dirac statistics. Accordingly, the wave function must be asymmetric under exchange of any pair of protons. Townes and Shawlow [64] showed that this constraint leads to an association of rotational states for which $k = 3n$, where $n = 0, 1, 2, \ldots$, with the 'parallel' nuclear spin state, $I = 3/2$ (ortho-NH$_3$). On the other hand, the rotational states with $k = 3n + 1, 3n + 2$ are associated with the 'anti-parallel' nuclear spin state, $I = 1/2$ (para-NH$_3$). Although the nuclear spin degeneracy of ortho-NH$_3$ ($2I + 1 = 4$) is twice that of para-NH$_3$ ($2I + 1 = 2$), the latter has twice as many rotational states (j, k): in any set of three consecutive values of k, two belong to para but only one to ortho. It follows that the total statistical weights of the ortho and the para levels are equal to 4. [Compare this with the case of H$_2$, where the total statistical weights of the ortho and para levels are 3 and 1, respectively.]

The internal rotational hamiltonian of a top may be written as

$$h = \frac{j_X^2}{2I_X} + \frac{j_Y^2}{2I_Y} + \frac{j_Z^2}{2I_Z} \qquad (4.84)$$

where j_X, j_Y, j_Z are the components of the rotational angular momentum, **j**, along the internal BF axes X, Y, Z; the Z-axis is taken to be the symmetry axis of the molecule. I_X, I_Y, I_Z are the moments of inertia along the BF axes. When the BF axes are taken to coincide with the principal axes of the molecule, the moment of inertia tensor, **I**, is diagonal:

$$\mathbf{I} = \begin{pmatrix} I_X & 0 & 0 \\ 0 & I_Y & 0 \\ 0 & 0 & I_Z \end{pmatrix}$$

In a symmetric top molecule, such as ammonia, $I_X = I_Y$. As $\mathbf{j}^2 = j_X^2 + j_Y^2 + j_Z^2$, the rotational hamiltonian takes the form

$$h = \frac{\mathbf{j}^2}{2I_X} + \left(\frac{1}{2I_Z} - \frac{1}{2I_X}\right) j_Z^2 \qquad (4.85)$$

The rotational eigenfunction, $|jkm\rangle$ [see equation (4.87) below], is an eigenfunction of $\mathbf{j}^2$ with eigenvalue $j(j+1)$, of j_Z with eigenvalue k, and of j_z with eigenvalue m (all in atomic units). Hence, we have the relation

$$h|jkm\rangle = \left[\frac{j(j+1)}{2I_X} + \left(\frac{1}{2I_Z} - \frac{1}{2I_X}\right) k^2\right] |jkm\rangle \qquad (4.86)$$

4.4 The rotational excitation of non-linear molecules

where the eigenvalues of the rotational hamiltonian are given by the term in square brackets in equation (4.86). To each value of the rotational quantum number, j, there correspond $(2j+1)$ values of the projection quantum number, $k = -j, -j+1, \ldots, 0, \ldots, j-1, j$. However, the expression for the eigenvalue involves k^2, and so the states $k = |k|$ and $k = -|k|$ are degenerate, i.e. they have the same eigenenergy.

The eigenfunctions, $|jkm\rangle$, of the rotational hamiltonian, h, are expressible in terms of the elements of the rotation matrix, $\mathbf{D}$ [cf. equation (4.12)], appropriately normalized, namely

$$|jkm\rangle = \left(\frac{2j+1}{8\pi^2}\right)^{-\frac{1}{2}} D^{j*}_{mk}(\alpha, \beta, \gamma) \tag{4.87}$$

where (α, β, γ) are the Euler angles that rotate the SF coordinate system into the BF system, defined above. As states $|jkm\rangle$ and $|j, -km\rangle$ are degenerate, the linear combinations

$$|jkm\epsilon\rangle = [2(1+\delta_{k0})]^{-\frac{1}{2}}(|jkm\rangle + \epsilon|j, -km\rangle) \tag{4.88}$$

with $\epsilon = \pm 1$ are also eigenfunctions of h with the same eigenenergy. The states with $\epsilon = \pm 1$ may be identified with the components of the *inversion doublets* which occur in NH$_3$, for $k > 0$, owing to the inversion motion of the nitrogen nucleus through the plane of the hydrogens. The state for which $(-1)^j \epsilon = +1$ is the asymmetric (upper) inversion state, whereas the state for which $(-1)^j \epsilon = -1$ is the symmetric (lower) inversion state. It may be seen from equation (4.88) that only the states with $\epsilon = +1$ exist when $k = 0$; this implies that the the lower (symmetric) inversion state is missing when $k = 0$ and j is even, and the upper (asymmetric) inversion state is missing when $k = 0$ and j is odd.

In practice, the inversion motion raises the degeneracy (i.e. gives rise to a splitting) of the states with $\epsilon = +1$ and $\epsilon = -1$. This splitting is small compared with the separation of the rotational levels and is neglected in the rigid rotor approximation. Nonetheless, this splitting is important spectroscopically: transitions between the two components of an inversion doublet ('inversion transitions') enable interstellar ammonia to be observed from the ground, at radio wavelengths.

The rotational excitation of NH$_3$ by He was studied by Green [65, 66]. We shall follow his approach, treating the molecule as a rigid symmetric top. In Fig. 4.3 the coordinates describing the interaction between NH$_3$ and He are shown schematically. The BF axes (X, Y, Z) provide a reference frame in which to locate the He atom, whose spherical polar coordinates are (R, θ', ϕ'). The interaction potential may be expanded in the form

$$V(R, \theta', \phi') = \sum_\lambda \sum_{\mu=-\lambda}^{\lambda} v_{\lambda\mu}(R) Y_{\lambda\mu}(\theta', \phi')$$

$$= \sum_\lambda \sum_{\mu=0}^{\lambda} v_{\lambda\mu}(R) \frac{Y_{\lambda\mu}(\theta', \phi') + (-1)^\mu Y_{\lambda,-\mu}(\theta', \phi')}{1 + \delta_{\mu 0}} \tag{4.89}$$

When writing (4.89), use is made of the facts that V is real and symmetric with respect to reflection in the XZ plane, and of the property

$$Y^*_{\lambda\mu}(\theta', \phi') = (-1)^\mu Y_{\lambda,-\mu}(\theta', \phi')$$

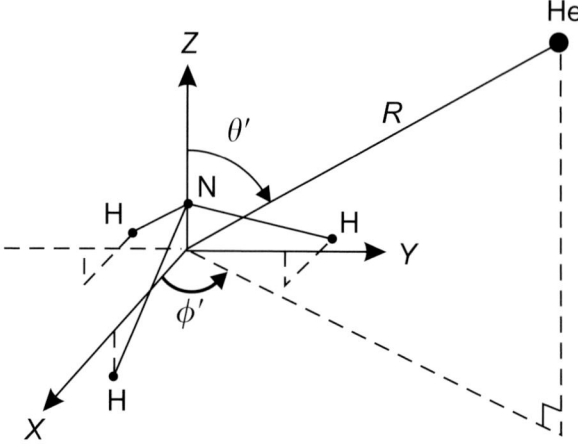

Figure 4.3 Defining the body-fixed coordinate system *XYZ* of NH$_3$, a symmetric top; the origin of the coordinate system is at the centre of mass of the molecule. The spherical polar coordinates (R, θ', ϕ'), determine the position of the He atom, relative to the body-fixed frame.

of the spherical harmonics. Owing to the three-fold symmetry of the potential, μ is restricted to integral multiples of 3.

The interaction potential (4.89) is expressed in the BF frame (X, Y, Z), whereas the rotational eigenfunctions, (4.87) and (4.88), are expressed in the SF frame (x, y, z). The relationship

$$Y_{\lambda\mu}(\theta', \phi') = \sum_{\nu} D^{\lambda}_{\nu\mu}(\alpha, \beta, \gamma) Y_{\lambda\nu}(\theta, \phi) \qquad (4.90)$$

[cf. equation (4.18)] transforms the spherical harmonics in (4.89) from the BF coordinates (θ', ϕ') into the SF coordinates (θ, ϕ), yielding $V(\alpha, \beta, \gamma, R, \theta, \phi)$. We recall that the Euler angles (α, β, γ) rotate the SF frame into the BF frame.

The total wave function describing the atom–molecule system may be expanded in terms of the functions of the total angular momentum, $\mathbf{J} = \mathbf{j} + \mathbf{l}$, where $\mathbf{j}$ is the rotational angular momentum of the molecule and $\mathbf{l}$ is the orbital angular momentum associated with the motion of the atom relative to the molecule:

$$|jk\epsilon lJM\rangle = \langle jmlm_l|JM\rangle |jkm\epsilon\rangle |lm_l\rangle \qquad (4.91)$$

In (4.91),

$$|lm_l\rangle = Y_{lm_l}(\theta, \phi)$$

and

$$\langle jmlm_l|JM\rangle = C^{jlJ}_{mm_lM}$$

is a Clebsch–Gordan coefficient; $|jkm\epsilon\rangle$ is given by equation (4.88). The total wave function

4.4 The rotational excitation of non-linear molecules

may be written in terms of the functions (4.91) as

$$\Psi(\alpha,\beta,\gamma,\mathbf{R}) = \sum_{\substack{jk\epsilon \\ lJM}} \frac{F(jk\epsilon lJM|R)}{R} |jk\epsilon lJM\rangle \quad (4.92)$$

where $\mathbf{R} = (R, \theta, \phi)$ is the position vector of the He atom in the SF coordinate system.

The Schrödinger equation (4.37) must now be solved. The total hamiltonian is

$$H = h - \frac{\nabla_R^2}{2\mu} + V(\alpha,\beta,\gamma,\mathbf{R}) \quad (4.93)$$

where h is the internal rotational hamiltonian (4.85). Schrödinger's equation reduces to

$$\left[\frac{d^2}{dR^2} - \frac{l(l+1)}{R^2} + \kappa_{jk}^2\right] F(jk\epsilon lJM|R)$$
$$= 2\mu \sum_{j'k'\epsilon'l'} \langle jk\epsilon lJM|V(\alpha,\beta,\gamma,\mathbf{R})|j'k'\epsilon'l'JM\rangle F(j'k'\epsilon'l'JM|R) \quad (4.94)$$

where

$$\kappa_{jk}^2 = 2\mu\left[E - \frac{j(j+1)}{2I_X} - \left(\frac{1}{2I_Z} - \frac{1}{2I_X}\right)k^2\right] \quad (4.95)$$

and we use κ, rather than k, to denote the wave number, in order to distinguish it from the projection of $\mathbf{j}$ on the symmetry axis of the molecule. Once again, the total angular momentum $\mathbf{J}$ and its projection M on the SF z-axis are conserved, i.e. their values remain unchanged during the collision.

The matrix elements of the interaction potential, V, which appear on the right-hand side of (4.94) may be derived as follows. First, we substitute equation (4.88) into equation (4.91) and obtain

$$|jk\epsilon lJM\rangle = \langle jmlm_l|JM\rangle$$
$$\times [2(1+\delta_{k0})]^{-\frac{1}{2}}(|jkm\rangle + \epsilon|j,-km\rangle)|lm_l\rangle$$
$$\equiv [2(1+\delta_{k0})]^{-\frac{1}{2}}(|jklJM\rangle + \epsilon|j,-klJM\rangle) \quad (4.96)$$

where

$$|jklJM\rangle = \langle jmlm_l|JM\rangle|jkm\rangle|lm_l\rangle \quad (4.97)$$

Second, we make use of the following relation, in terms of 3j- and 6j-coefficients,

$$\langle jklJM|V(\alpha,\beta,\gamma,\mathbf{R})|j'k'l'JM\rangle$$
$$= \sum_{\lambda\mu} v_{\lambda\mu}(-1)^{j+j'+k'-J}\left[\frac{(2j+1)(2j'+1)(2l+1)(2l'+1)(2\lambda+1)}{4\pi}\right]^{\frac{1}{2}}$$
$$\times \begin{pmatrix} l & l' & \lambda \\ 0 & 0 & 0 \end{pmatrix}\begin{pmatrix} j & j' & \lambda \\ k & -k' & \mu \end{pmatrix}\begin{Bmatrix} j' & l' & J \\ l & j & \lambda \end{Bmatrix} \quad (4.98)$$

which expresses the matrix elements of V in the basis (4.97). We note that equation (4.98) is independent of M, the projection of the total angular momentum, **J**, on the z-axis.

The coupled equations (4.94) must be solved, subject to the appropriate physical boundary conditions, to obtain the reactance matrix, **K**, and hence the scattering matrix, **S**. The partial cross-sections are then derived from

$$\sigma_J(jk\epsilon \leftarrow j'k'\epsilon') = \frac{\pi}{\kappa_{j'k'}^2(2j'+1)}$$

$$\times \sum_{l,l'} (2J+1)|T_J(jk\epsilon l, j'k'\epsilon'l')|^2 \quad (4.99)$$

where $T_J(jk\epsilon l, j'k'\epsilon'l')$ is an element of the *transmission* matrix, which is related to the **S** matrix by

$$\mathbf{T} = \mathbf{1} - \mathbf{S} \quad (4.100)$$

The total cross-section is obtained by summing (4.99) over J.

It will be recalled that the index μ, which appears in the expansion (4.89), is restricted to integral multiples of 3, owing to the three-fold symmetry of the interaction potential about the symmetry axis of the ammonia molecule. As a consequence, the 3j-symbol

$$\begin{pmatrix} j & j' & \lambda \\ k & -k' & \mu \end{pmatrix}$$

in (4.98) restricts changes in the projection quantum number to $\Delta k = 3n$, where n is an integer. This restriction relates to the fact that ortho- and para-NH$_3$ may be treated as distinct species, as far as non-reactive collisions are concerned. Just as in the case of H$_2$, the ortho and para forms interconvert only through proton-exchange reactions.

4.4.2 Asymmetric tops

The category of asymmetric tops comprises important interstellar molecules such as water (H$_2$O) and formaldehyde (H$_2$CO). Some polyatomic molecules, such as methanol (CH$_3$OH), are asymmetric tops that are close to being symmetric tops [if the internal rotation of the methyl (CH$_3$) relative to the hydroxyl (OH) group – the *torsional motion*, which is analogous to a vibration – is neglected].

In an asymmetric top, the three principal moments of inertia have different values, i.e. $I_X \neq I_Y \neq I_Z$, and the internal rotational hamiltonian (4.84) may be written as

$$h = \frac{\mathbf{j}^2}{2I_X} + \left(\frac{1}{2I_Y} - \frac{1}{2I_X}\right)j_Y^2 + \left(\frac{1}{2I_Z} - \frac{1}{2I_X}\right)j_Z^2$$

$$\equiv A\mathbf{j}^2 + (B-A)j_Y^2 + (C-A)j_Z^2 \quad (4.101)$$

where $A = 1/(2I_X)$, $B = 1/(2I_Y)$ and $C = 1/(2I_Z)$ are the rotational constants. It follows that

$$h|jkm\rangle = [Aj(j+1) + (C-A)k^2 + (B-A)j_Y^2]|jkm\rangle \quad (4.102)$$

where $|jkm\rangle$ is given by equation (4.87).

4.4 The rotational excitation of non-linear molecules

The term $(B - A)j_Y^2$ in the hamiltonian h determines the degree of asymmetry of the top. The step-up, j_+, and step-down, j_-, angular momentum (ladder) operators are defined by

$$j_+ = j_X + ij_Y$$

and

$$j_- = j_X - ij_Y$$

whence

$$2ij_Y = j_+ - j_-$$

and

$$-4j_Y^2 = j_+^2 + j_-^2 - j_+j_- - j_-j_+$$

Operating on $|jkm\rangle$ with j_Y^2, we obtain

$$j_Y^2|jkm\rangle = -\frac{1}{4}\left\{[(j-k)(j+k+1)(j-k-1)(j+k+2)]^{\frac{1}{2}}|j,k+2,m\rangle \right.$$
$$+ [(j+k)(j-k+1)(j+k-1)(j-k+2)]^{\frac{1}{2}}|j,k-2,m\rangle$$
$$\left. -2\left[j(j+1) - k^2\right]\right\}|jkm\rangle \tag{4.103}$$

and equation (4.102) becomes

$$h|jkm\rangle = \left[\frac{(A+B)}{2}j(j+1) + \left(C - \frac{(A+B)}{2}\right)k^2\right]|jkm\rangle$$
$$+ \frac{(A-B)}{4}[(j-k)(j+k+1)(j-k-1)(j+k+2)]^{\frac{1}{2}}|j,k+2,m\rangle$$
$$+ \frac{(A-B)}{4}[(j+k)(j-k+1)(j+k-1)(j-k+2)]^{\frac{1}{2}}|j,k-2,m\rangle$$
$$\equiv \frac{(A+B)}{2}[j(j+1) - k^2]|jkm\rangle + Ck^2|jkm\rangle$$
$$+ \frac{(A-B)}{4}[j(j+1) - k(k+1)]^{\frac{1}{2}}$$
$$\times [j(j+1) - (k+1)(k+2)]^{\frac{1}{2}}|j,k+2,m\rangle$$
$$+ \frac{(A-B)}{4}[j(j+1) - k(k-1)]^{\frac{1}{2}}$$
$$\times [j(j+1) - (k-1)(k-2)]^{\frac{1}{2}}|j,k-2,m\rangle \tag{4.104}$$

We see from equation (4.104) that $|jkm\rangle$ is *not* an eigenfunction of h, as $|j, k \pm 2, m\rangle$ appear on the right-hand side. Put another way, the matrix of h in the basis $|jkm\rangle$ is not diagonal in k: there are off-diagonal elements involving $k \pm 2$. The relative magnitudes of the off-diagonal and diagonal elements involving k are proportional to $(A - B)/(2C - A - B)$. We note that the off-diagonal elements vanish in the limit of the symmetric top, $A = B$, as must be the case.

The eigenenergies of the asymmetric top may be obtained by diagonalizing the hamiltonian matrix. The projection, k, is no longer a 'good' quantum number. As the off-diagonal couplings are to states with $k \pm 2$, either even or odd values of k are 'mixed' through the diagonalization procedure. In other words, the eigenfunctions of h may be written as linear combinations of the eigenfunctions of the symmetric top, $|jkm\rangle$, for given values of j and m,

$$|j\tau m\rangle = \sum_k a_{\tau k} |jkm\rangle \qquad (4.105)$$

where the sum extends over either even or odd values of k, and τ labels the asymmetric top eigenfunctions. The number of the asymmetric top eigenfunctions is the same as the number of the 'primitive' symmetric top functions of which they are composed; τ is taken to be an integer, $-j \leq \tau \leq j$, which orders the energy levels for a given value of j. In the case of a near-symmetric top, $A \approx B$, a particular value of k dominates the expansion on the right-hand side of equation (4.105), i.e. the corresponding value of $a_{\tau k} \approx 1$, whereas $a_{\tau k} \ll 1$ for all other values of k. The label τ may then be identified with the dominant value of k. In the limit of the symmetric top, $A = B$. If $A = B > C$, the top is *oblate*, whereas, if $A = B < C$, the top is *prolate*.

The first quantitative calculations for an astrophysically important asymmetric top molecule related to formaldehyde (H$_2$CO), in collision with He [67]. The formaldehyde molecule is shown schematically in Fig. 4.4. The atoms comprising the molecule all lie in the XZ plane, with H atoms symmetrically disposed either side of the symmetry axis, i.e. the H atoms have the same Z coordinate and their X coordinates are equal in magnitude but opposite in sign. We have already noted [equation (4.87)] that the basis functions of the asymmetric top are the normalized rotation matrix elements

$$|jkm\rangle = \left(\frac{2j+1}{8\pi^2}\right)^{\frac{1}{2}} D^{j*}_{mk}(\alpha, \beta, \gamma)$$

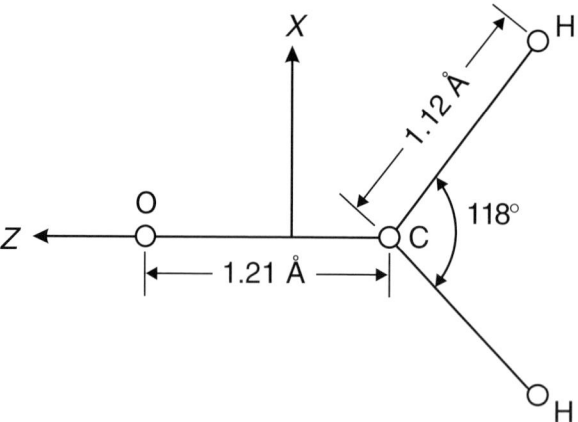

Figure 4.4 Schematic diagram of the formaldehyde molecule. The coordinate origin is located at the centre of mass of the molecule. Distances are in units of 10^{-10} m.

4.4 The rotational excitation of non-linear molecules

Exchange of the (identical) hydrogen nuclei is effected by the transformation $\gamma \to \gamma + \pi$, under which $|jkm\rangle \to \exp(ik\pi)|jkm\rangle$. Thus, if k is even, $|jkm\rangle \to |jkm\rangle$, whereas, if k is odd, $|jkm\rangle \to -|jkm\rangle$, i.e. the basis functions are symmetric or asymmetric under exchange of the H nuclei, according as k is even or odd, respectively. As the H nuclei (protons) are fermions, their total wave function, including the spin function, must be asymmetric under their exchange. The spin functions are

$$I = 1, M_I = 1: \quad \alpha_1 \alpha_2$$
$$I = 1, M_I = 0: \quad (\alpha_1 \beta_2 + \alpha_2 \beta_1)/2^{\frac{1}{2}}$$
$$I = 1, M_I = -1: \quad \beta_1 \beta_2$$

and

$$I = 0, M_I = 0: \quad (\alpha_1 \beta_2 - \alpha_2 \beta_1)/2^{\frac{1}{2}}$$

for the ortho ($I = 1$) and the para ($I = 0$) states; M_I is the corresponding projection quantum number. α denotes the proton spin state with projection $m_s = 1/2$ ('spin up') and β the state with $m_s = -1/2$ ('spin down'), and the subscripts '1' and '2' label the two protons. The ortho states are symmetric under proton exchange, i.e. under interchange of the subscripts '1' and '2', whereas the para state is asymmetric. As the total wave function is a product of the spin and the rotation parts, Fermi–Dirac statistics require that ortho-H_2CO is associated with rotational states for which k is odd only, and para-H_2CO is associated with rotational states for which k is even only. (This circumstance is similar to that in H_2, where j is odd in ortho-H_2 and even in para-H_2.) Transitions between the ortho and the para forms can be effected only by proton or hydrogen exchange reactions.

The rotational constants of formaldehyde are $A = 1.295\,\text{cm}^{-1}$, $B = 1.134\,\text{cm}^{-1}$, $C = 9.407\,\text{cm}^{-1}$, where A, B and C relate to the X, Y, and Z axes, defined in Fig. 4.4. When classifying the degree of asymmetry of tops, it is conventional to order the rotational constants such that $A > B > C$. In the example of formaldehyde, considered above, this is equivalent to a cyclic permutation of the coordinate axes, $X \to Y \to Z \to X$. The *Ray asymmetry parameter* is then defined as

$$\kappa = \frac{2B - A - C}{A - C}$$

[64]. In the limit of a prolate symmetric top ($B = C$), $\kappa = -1$, and in the limit of an oblate symmetric top ($B = A$), $\kappa = 1$. Taking the values of the rotational constants that correspond to formaldehyde, we obtain $\kappa = -0.961$, which is not far from the prolate symmetric top limit. Water (H_2O), on the other hand, for which $\kappa = -0.436$, is further from this limit.

The spectroscopic notation for the energy levels of asymmetric tops is often given in terms of the values of the projection quantum number k in the prolate and oblate symmetric top limits, k_{-1} and k_{+1} (or k_- and k_+), respectively. The index $\tau = k_{-1} - k_{+1}$ [64]. The energy levels of H_2O (para and ortho) up to 200 cm^{-1} above the ground state of para-H_2O are listed in Table 4.1. The energy levels may be labelled by j_τ or by $j_{k_- k_+}$.

The rotational excitation of molecules

Table 4.1. *Energy levels of para- and ortho-*H_2O *up to* $200\,\text{cm}^{-1}$ *above the ground state of para-*H_2O *(from [68]). The alternative methods of labelling the energy levels, by* j_τ *or by* $j_{k_-k_+}$, *are given.*

j	τ	k_-	k_+	Energy (cm^{-1})	modification
0	0	0	0	0.0000	para
1	−1	0	1	23.7943	ortho
1	0	1	1	37.1371	para
1	1	1	0	42.3717	ortho
2	−2	0	2	70.0907	para
2	−1	1	2	79.4963	ortho
2	0	1	1	95.1757	para
2	1	2	1	134.9018	ortho
2	2	2	0	136.1641	para
3	−3	0	3	136.7617	ortho
3	−2	1	3	142.2783	para
3	−1	1	2	173.3656	ortho

4.4.3 Asymmetric tops with internal rotation

The example that we shall take of molecules in this category is methanol (CH_3OH). Methanol is one of the most important interstellar molecules. It has been observed in many millimetre and sub-millimetre transitions, both in dark molecular clouds, where its abundance, relative to H_2, is of the order of 10^{-9}, and in high mass protostellar objects, where its fractional abundance can reach 10^{-6}. Methanol is both a maser (in protostellar objects) and is sometimes observed in absorption against the cosmic background (2.73 K) radiation field (in dark clouds). Grain-surface reactions are believed to be important in the production of methanol, which can be released into the gas phase by sublimation processes or by sputtering, induced by shock waves.

The richness of the spectrum of methanol reflects the complexity of its internal structure. In Fig. 4.5 is shown the equilibrium configuration of CH_3OH. The hydrogens of the methyl (CH_3) group form an equilateral triangle, as in the case of NH_3. The COH group defines a plane containing the BF Z-axis, which is perpendicular to the plane of the hydrogens and passes through the centre of mass of the molecule. The symmetry axis of the methyl group is perpendicular to the plane of the hydrogens and passes through its geometrical centre; it is parallel to, but slightly displaced from, the Z-axis.

Methanol is a slightly asymmetric top, with rotational constants $A = 0.823\,\text{cm}^{-1}$, $B = 0.793\,\text{cm}^{-1}$, and $C = 4.257\,\text{cm}^{-1}$ along the X, Y and Z axes (cf. Fig. 4.5), respectively, in its lowest torsional state [70]. Reordering the rotational constants according to the convention that $A > B > C$, which is achieved by a cyclic permutation of the coordinate axes, $X \rightarrow Y \rightarrow Z \rightarrow X$, the Ray asymmetry parameter takes the value $\kappa = -0.983$, which is close to the prolate symmetric top limit (-1). The three protons of the CH_3 group may have 'parallel' or 'anti-parallel' spins, and these spin states can interconvert only through proton exchange reactions, just as in the case of ammonia. In A-type methanol, the total spin is $I = 3/2$, whereas, in E-type methanol, $I = 1/2$; these are equivalent to the ortho and para forms, respectively, of ammonia. The nuclear spin degeneracy, $(2I + 1)$, is 4 for A-type and

4.4 The rotational excitation of non-linear molecules

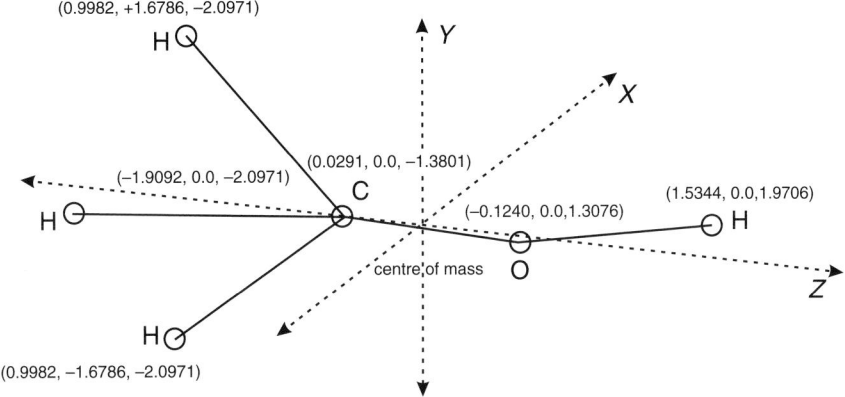

Figure 4.5 Defining the body-fixed coordinate system of methanol (CH$_3$OH), a near symmetric top. The (X, Y, Z) coordinates of the atoms correspond to the equilibrium configuration of the molecule [69]. Methanol exhibits internal torsional motion of the CH$_3$ group relative to the OH group.

2 for E-type. However, E-type methanol has two degenerate forms, E$_1$ and E$_2$, and hence the total number of E-type states is equal to the total number of A-type states [71]. Our treatment of the internal motion of methanol closely follows that of Lin and Swalen [72]; but it should be noted that Lin and Swalen took the YZ plane to be the plane of symmetry of the molecule, whereas we adopt the XZ plane (cf. Fig. 4.5).

Let us denote the eigenfunctions of methanol by $|jkv\sigma\rangle$, where, as previously, j and k denote the rotational angular momentum and its projection on the symmetry (Z-) axis; v denotes the torsional state, and $\sigma = 0$ for A-type methanol, $\sigma = 1$ for E$_1$-type and $\sigma = -1$ for E$_2$-type. Then the non-vanishing matrix elements of the internal hamiltonian, h, may be written

$$\langle jkv\sigma | h | jkv\sigma \rangle$$
$$= \frac{(A+B)}{2}[j(j+1) - k^2] + Ck^2 + E_{kv\sigma} \quad (4.106)$$

$$\langle jkv\sigma | h | j, k+2, v'\sigma \rangle$$
$$= \frac{(A-B)}{4}[j(j+1) - k(k+1)]^{\frac{1}{2}}$$
$$\times [j(j+1) - (k+1)(k+2)]^{\frac{1}{2}} I_{kv\sigma}^{k+2,v'\sigma} \quad (4.107)$$

$$\langle jkv\sigma | h | j, k-2, v'\sigma \rangle$$
$$= \frac{(A-B)}{4}[j(j+1) - k(k-1)]^{\frac{1}{2}}$$
$$\times [j(j+1) - (k-1)(k-2)]^{\frac{1}{2}} I_{kv\sigma}^{k-2,v'\sigma} \quad (4.108)$$

$$\langle jk\nu\sigma | h | j, k+1, \nu'\sigma \rangle$$
$$= \frac{D}{2}(2k+1)[j(j+1) - k(k+1)]^{\frac{1}{2}} I_{k\nu\sigma}^{k+1,\nu'\sigma} \tag{4.109}$$

and

$$\langle jk\nu\sigma | h | j, k-1, \nu'\sigma \rangle$$
$$= \frac{D}{2}(2k+1)[j(j+1) - k(k-1)]^{\frac{1}{2}} I_{k\nu\sigma}^{k-1,\nu'\sigma} \tag{4.110}$$

$E_{k\nu\sigma} \equiv E_{-k\nu-\sigma}$ is the eigenvalue of the equation for the torsional motion, and $I_{k\nu\sigma}^{k'\nu'\sigma}$ is an overlap matrix element of the torsional eigenfunctions:

$$I_{k\nu\sigma}^{k'\nu'\sigma} = \langle k\nu\sigma | k'\nu'\sigma \rangle$$

The torsional motion is not free, but 'hindered' by a three-fold symmetric potential

$$V(\omega) = \frac{V_3}{2}(1 - \cos 3\omega) \tag{4.111}$$

where ω is the angle of rotation of the methyl, relative to the hydroxyl group; $V_3 = 373 \text{ cm}^{-1}$ is the height of the barrier to torsional motion in methanol. In the limit of an infinite barrier to torsional motion, $I_{k\nu\sigma}^{k'\nu'\sigma} = \delta_{\nu\nu'}$.

Equations (4.106–4.110) are generalizations of the previous results for the rigid asymmetric top. The additional matrix elements (4.109, 4.110) arise from the lack of rigidity of methanol, which gives rise to its torsional motion. The additional constant $D = 0.0026 \text{ cm}^{-1} \ll A, B, C$.

The energy associated with the torsional motion, $E_{k\nu\sigma}$, is much larger for A-type than for E-type methanol. As these energies appear along the diagonal of the internal hamiltonian matrix, it follows that the off-diagonal terms are smaller, relative to those on the diagonal, in A-type. In the limit of vanishingly small off-diagonal terms, the hamiltonian reduces to that of the symmetric top. This limit is approached for large j and k in both A-type and E-type, as examination of the off-diagonal terms (4.107–4.110) confirms. In the case of A-type methanol, the symmetric top limit is approached also for small j and k, and hence states $|j, \pm k, \nu, \sigma = 0\rangle$ are almost degenerate.

The eigenfunctions of the torsional hamiltonian, $|k\nu\sigma\rangle$, may be written as linear combinations of free rotor basis functions, $|km_\omega\rangle$, where m_ω denotes the torsional angular momentum of the methyl group,

$$|k\nu\sigma\rangle = \sum_{m_\omega = 3s + \sigma} a_{k\nu m_\omega} |km_\omega\rangle, \tag{4.112}$$

where s is an integer and

$$|km_\omega\rangle = (2\pi)^{-\frac{1}{2}} e^{-i\rho k\omega} e^{im_\omega \omega}, \tag{4.113}$$

ρ is a dimensionless molecule-specific parameter [72], given by

$$\rho = I_\omega / I_z,$$

4.4 The rotational excitation of non-linear molecules

where I_ω is the moment of inertia of the methyl group and I_Z is the moment of inertia of the whole molecule about the Z-axis; the numerical value appropriate to methanol is $\rho = 0.8097$ [73].

The torsional wave function, $M(\omega) \equiv |kv\sigma\rangle$, is the solution of the *Mathieu equation*

$$\frac{1}{\alpha}\frac{d^2 M(\omega)}{d\omega^2} + (R + 2\cos 3\omega)M(\omega) = 0 \tag{4.114}$$

where

$$\alpha = V_3/4F$$

and

$$R = \frac{4}{V_3}\left(W - \frac{1}{2}V_3\right)$$

$W \equiv E_{kv\sigma}$ is the torsional eigenenergy and F is the torsional constant. In the case of methanol, $F = 27.633$ cm^{-1} and $V_3 = 373.1$ cm^{-1} [70]. The torsional quantum number, v, labels the different eigenvalues that have the same value of k.

In the case of $\sigma = 0$ (A-type methanol) and $k = 0$, the solutions to the Mathieu equation are analogous to the symmetrized wave functions of the simple harmonic oscillator,

$$M(\omega) = (2\pi)^{-\frac{1}{2}} \sum_{s \geq 0} (a_s e^{i3s\omega} + a_{-s} e^{-i3s\omega}), \tag{4.115}$$

in which $a_s = a_{-s}$ or $a_s = -a_{-s}$, corresponding to the torsional wave function being either symmetric or antisymmetric about the equilibrium position, $\omega = 0$, and v being even or odd, respectively. The index v is analogous to that used to label the eigenfunctions of the simple harmonic oscillator. The lowest energy state for a given k has the quantum number $v = 0$, as for the simple harmonic oscillator. In general, the solutions of the Mathieu equation have to be obtained numerically.

The lowest state of E-type methanol, $(j = 1, k = -1)$, lies slightly higher (by 5.49 cm^{-1}) than the lowest state of A-type, $(j = 0 = k)$. This energy difference is analogous to, although much smaller than, that between the lowest states of ortho- and para-H$_2$, where the $j = 1$ ortho state is higher, by 118 cm^{-1}, than the $j = 0$ para state. Such shifts are without consequence in the context of *non*-reactive scattering, as the E/A types or para/ortho forms may be treated as distinct species that cannot interconvert. However, proton exchange reactions with ions such as H$^+$ and H$_3^+$ can interconvert these species, and the energy shifts may become significant, in this context. In the case of methanol, the energy shift is small (approximately 8 K) and has an effect on the relative abundances of E- and A-type only in very cold molecular clouds.

5
The vibrational excitation of linear molecules

5.1 Introduction

In this chapter, we extend the treatment of rotational excitation, presented in Chapter 4, to encompass vibrational transitions. In practice, the discussion will be limited to linear molecules for which calculations are feasible and quantitative results have been obtained. As compared with rotational transitions, which may be excited at kinetic temperatures $T \approx 10$ K, in the case of heavy molecules, or $T \approx 100$ K, in the case of light (hydrogen-bearing) molecules, vibrational excitation generally requires $T \approx 1000$ K. There are exceptions: the inversion motion (of the nitrogen in the plane of the hydrogen atoms) in NH_3 is a form of vibrational motion, and the energy involved is small, of the order of 1 K. Similarly, the torsional motion (of the CH_3 relative to the OH group) in CH_3OH may be viewed as another form of vibrational motion and involves energies of the order of 100 K. These phenomena have been discussed in Chapter 4. In the present chapter, we shall be concerned with the stretching of chemical bonds within molecules, a process that involves higher energies, on the order of 1000 K.

Vibrational transitions are important in a number of astrophysical contexts, including cool stellar atmospheres, shocked regions of molecular clouds, and planetary nebulae. Vibrational transitions occur between specific rovibrational states (v, j) of the molecule. Molecular hydrogen has rovibrational transitions in the near infrared, at about 2 μm, some of which can be observed from the ground through atmospheric windows. Similarly, CO has near-infrared rovibrational transitions which are observable from the ground. Rovibrational emission may be excited by radiative pumping or by collisions with the abundant species, H_2, H and He. The following discussion relates to collisional excitation; radiative transitions are considered in Chapter 10.

5.2 The scattering of an atom by a vibrating rotor

Two extensions of the theory presented in the previous chapter are necessary in order to treat the excitation of a vibrating rotor by an atom. First, the additional internal degree of freedom of the diatom BC, represented by the internuclear distance, r, must be taken into account; r will be referred to as the *vibrational coordinate*. Second, the dependence on r on the interaction potential, V, must be known and expressible in the form

$$V(r, R, \theta') = \sum_{\lambda=0}^{\infty} v_\lambda(r, R) P_\lambda(\cos \theta') \qquad (5.1)$$

5.2 The scattering of an atom by a vibrating rotor

where the coordinates are defined in Fig. 4.2; r is the distance BC in this figure. The summation in equation (5.1) might (depending on the degree of anisotropy of the potential) be restricted to a small number of terms.

The Schrödinger equation

$$H\Psi = E\Psi \tag{5.2}$$

can be solved using either space-fixed (SF) coordinates, in which case $\Psi = \Psi(\mathbf{r}, \mathbf{R})$, or body-fixed (BF) coordinates, where $\Psi = \Psi(\mathbf{r}', \mathbf{R})$. We note that $|\mathbf{r}| = |\mathbf{r}'|$: the choice of coordinate system affects the rotational but not the vibrational coordinate (see below).

The hamiltonian, H, of the atom–molecule system may be written as

$$H = h - \frac{\nabla_R^2}{2\mu} + V(r, R, \theta') \tag{5.3}$$

where

$$h = -\frac{\nabla_r^2}{2m} + v(r) \tag{5.4}$$

is the internal hamiltonian of the diatomic molecule BC at infinite separation from the atom, A; $m = m_B m_C/(m_B + m_C)$ is the reduced mass of the molecule. The kinetic energy operator in equation (5.4) may be separated into its radial and angular parts,

$$-\frac{1}{2m}\nabla_r^2 = -\frac{1}{2mr}\frac{\partial^2}{\partial r^2}r + \frac{\mathbf{j}^2}{2mr^2} \tag{5.5}$$

and we recall that $\mathbf{j}$ is the rotational angular momentum of the molecule.

Let us now consider the eigenvalue equation for the isolated vibrating rotor:

$$h\psi(\mathbf{r}) = \epsilon\psi(\mathbf{r}) \tag{5.6}$$

The form that will be adopted for the vibrational eigenfunction is

$$\psi(\mathbf{r}) = \frac{\chi(vj|r)}{r} Y_{jm_j}(\hat{\mathbf{r}}) \tag{5.7}$$

where v is the vibrational quantum number.

The vibrational wave function, $\chi(vj|r)$, satisfies the differential equation

$$\left[-\frac{d^2}{dr^2} + \frac{j(j+1)}{r^2} + 2mv(r) - \kappa^2 \right] \chi(vj|r) = 0 \tag{5.8}$$

where $\kappa^2 = 2m\epsilon$. We note that equation (5.8) is independent of the projection quantum number, m_j.

Equation (5.8) can be solved numerically, subject to the appropriate boundary conditions,

$$\chi(vj|r) \to 0$$

as $r \to 0$ or $r \to \infty$. Alternatively, approximate solutions may be sought. A frequently used approximation is to replace r^2 by its expectation value in the lowest rotational state, $j = 0$:

$$\langle r^2 \rangle_v = \int \chi^*(v0|r) r^2 \chi(v0|r) dr \tag{5.9}$$

84   *The vibrational excitation of linear molecules*

This approximation neglects the centrifugal stretching of $\langle r^2 \rangle$, owing to the rotational motion. Substituting (5.9) into (5.8) leads to

$$\left[-\frac{d^2}{dr^2} + 2mv(r) - \kappa_v^2 \right] \chi(vj|r) = 0 \tag{5.10}$$

where

$$\kappa_v^2 = 2m\left[\epsilon - \frac{j(j+1)}{2m\langle r^2 \rangle_v} \right]$$

$$= 2m[\epsilon - B_v j(j+1)] \tag{5.11}$$

and where

$$B_v = 1/(2m\langle r^2 \rangle_v) \tag{5.12}$$

is the rotational constant in the vibrational state v. The term in square brackets on the right-hand side of equation (5.11) is the difference between the rovibrational energy ϵ of the molecule BC and the rotational energy $B_v j(j+1)$, i.e. the energy of the lowest level $v, j = 0$ of each vibrational manifold.

If small values of the vibrational quantum number, v, are relevant – which is often the case – then a harmonic oscillator potential might be a satisfactory approximation to $v(r)$, that is

$$v(r) = \frac{1}{2} K (r - r_0)^2 \tag{5.13}$$

where the *force constant*, K, and the *equilibrium distance*, r_0, are determined by a best fit to the energy levels, as determined from spectroscopic data. An alternative approach, which is often preferable, is to adopt a Morse oscillator potential

$$v(r) = D \left[e^{-a(r-r_0)} - 1 \right]^2 \tag{5.14}$$

in which a and D are the constants to be derived by a best fit to the spectroscopic data; D may be identified with the dissociation energy of the molecule, measured relative to the minimum of the potential well, $r = r_0$, where $v(r) = 0$. By analogy with equations (4.35) and (4.36), the total wave function, Ψ, of the atom–molecule system may be expanded in terms of functions of radial and angular coordinates. Thus, if SF coordinates are used

$$\Psi(\mathbf{r}, \mathbf{R}) = \sum_{vjlpJM} \frac{F(vjlpJM|R)}{R} \frac{\chi(vj|r)}{r} \mathcal{Y}_{jlpJM}(\hat{\mathbf{r}}; \hat{\mathbf{R}}) \tag{5.15}$$

with $\mathcal{Y}_{jlpJM}(\hat{\mathbf{r}}; \hat{\mathbf{R}})$ defined as in Chapter 4. Similarly, in terms of BF coordinates

$$\Psi(\mathbf{r}', \mathbf{R}) = \sum_{vj\bar{\Omega}pJM} \frac{G(vj\bar{\Omega}pJM|R)}{R} \frac{\chi(vj|r)}{r} \mathcal{Z}_{j\bar{\Omega}pJM}(\hat{\mathbf{r}}'; \hat{\mathbf{R}}) \tag{5.16}$$

5.2 The scattering of an atom by a vibrating rotor

where $\mathcal{Z}_{j\bar{\Omega}pJM}(\hat{\mathbf{r}}';\hat{\mathbf{R}})$ is also defined in Chapter 4. Using techniques analogous to those employed in Chapter 4, the Schrödinger equation (5.2) may be reduced to a set of coupled differential equations,

$$\left[\frac{d^2}{dR^2} - \frac{l(l+1)}{R^2} + k_{vj}^2\right] F(vjlpJ|R)$$
$$= 2\mu \sum_{v'j'l'\lambda} f_\lambda(jl, j'l'; J) y_\lambda(vj, v'j'|R) F(v'j'l'pJ|R) \quad (5.17)$$

where $f_\lambda(jl, j'l'; J)$ is a Percival–Seaton coefficient, defined by equation (4.56), and

$$y_\lambda(vj, v'j'|R) = \int_0^\infty \chi^*(vj|r) v_\lambda(r, R) \chi(v'j'|r) dr \quad (5.18)$$

determines the coupling between the vibrational states v and v'; $k_{vj}^2 = 2\mu(E - \epsilon_{vj})$, where $\epsilon_{vj} = \kappa_v^2/2m + B_v j(j+1)$ is the rovibrational energy of the molecule BC in the state (v, j). On the other hand, the BF expansion (5.16) leads to

$$\left[\frac{d^2}{dR^2} + k_{vj}^2\right] G(vj\bar{\Omega}pJ|R)$$
$$= 2\mu \sum_{v'j'\bar{\Omega}'} V_{\text{eff}}(vj\bar{\Omega}, v'j'\bar{\Omega}'; J|R) G(v'j'\bar{\Omega}'pJ|R) \quad (5.19)$$

where

$$V_{\text{eff}}(vj\bar{\Omega}, v'j'\bar{\Omega}'; J|R)$$
$$= \langle vj\bar{\Omega}pJM | V(r, R, \theta') + \frac{\mathbf{l}^2}{2\mu R^2} | v'j'\bar{\Omega}'pJM \rangle \quad (5.20)$$

are the matrix elements of the effective potential

$$V_{\text{eff}}(r, R, \theta') = V(r, R, \theta') + \frac{\mathbf{l}^2}{2\mu R^2} \quad (5.21)$$

and are independent of the projection quantum number, M. The non-vanishing elements of the centrifugal potential operator, $\mathbf{l}^2/(2\mu R^2)$, are given by equations (4.63) and (4.64). By analogy with equation (4.62), the non-vanishing elements of the interaction potential, $V(r, R, \theta')$, are

$$\langle vj\bar{\Omega}pJM | V(r, R, \theta') | v'j'\bar{\Omega}pJM \rangle$$
$$= (-1)^{\bar{\Omega}} \sum_\lambda \frac{[(2j+1)(2j'+1)]^{\frac{1}{2}}}{(2\lambda+1)} C_{000}^{jj'\lambda} C_{\bar{\Omega},-\bar{\Omega}0}^{jj'\lambda} y_\lambda(vj, v'j'|R) \quad (5.22)$$

where $y_\lambda(vj, v'j'|R)$ is defined by equation (5.18) above.

Only for the lightest of molecules, which have the largest values of the rotational constant, B_v, is it feasible to undertake coupled-channel calculations involving the solution of the full

86   *The vibrational excitation of linear molecules*

coupled-channel (CC) equations (5.17) or (5.19). The reason is readily seen: each vibrational manifold, v, contains a certain number, j_v, of bound rotational states. In a CC calculation, the numbers of coupled channels are

$$\sum_{j=0}^{j_v}(j+1) = (j_v+1)(j_v+2)/2$$

for one value of the parity, p, and

$$\sum_{j=0}^{j_v} j = j_v(j_v+1)/2$$

for the other parity. Taking $j_v = 20$, for example, there are 231 and 210 coupled equations, respectively, for the vibrational manifold, v. In practice, it may be necessary to consider several vibrational manifolds simultaneously. Furthermore, the energy levels of the vibrational manifolds overlap, as states with large j and small v have similar energies to states with small j and large v. As it is usually necessary, in CC calculations, to include all states below a given energy in the basis, for both completeness and reliability of the results, large values of j_v are necessary for the lower manifolds. The quadratic dependence of the number of coupled channels on j_v and the coupling between manifolds ensure that the problem rapidly becomes very computationally demanding. With the exception of the most favourable molecule, H_2 (and its deuterated isotope, HD), for which CC calculations are feasible, approximate methods still have to be adopted. We discuss first the approximate methods, and then consider the specific examples of rovibrational excitation of H_2 and HD by H and He.

5.2.1   *Decoupling approximations*

For studies of the vibrational excitation of light neutral molecules by neutral perturbers, angular momentum decoupling, as in the centrifugal decoupling [coupled states (CS)] approximation considered in Chapter 4, is appropriate. This expectation has been confirmed in the case of H_2–He collisions, for example. When the same H_2–He interaction potential is employed, good agreement is found between the results of CC and CS calculations; cross-sections for $(v',j') \rightarrow (v,j) = (0,j)$ are shown in Fig. 5.1. The level of agreement is all the more striking in view of the rapid variation of the cross-sections with the collision energy as the threshold is approached.

For heavy neutral molecules, the energy sudden approximation may be used in conjunction with the centrifugal decoupling approximation. The resulting infinite-order sudden (IOS) approximation has been used extensively in calculations of rovibrational excitation cross-sections. Following the analysis in Chapter 4, it may readily be shown that, in this case, the coupled equations (5.19) reduce to the form

$$\left[\frac{d^2}{dR^2} - \frac{l(l+1)}{R^2} + k_v^2\right] G_l(v|R,\theta')$$
$$= 2\mu \sum_{v'\lambda} P_\lambda(\cos\theta') y_\lambda(v,v'|R) G_l(v'|R,\theta') \quad (5.23)$$

5.2 The scattering of an atom by a vibrating rotor

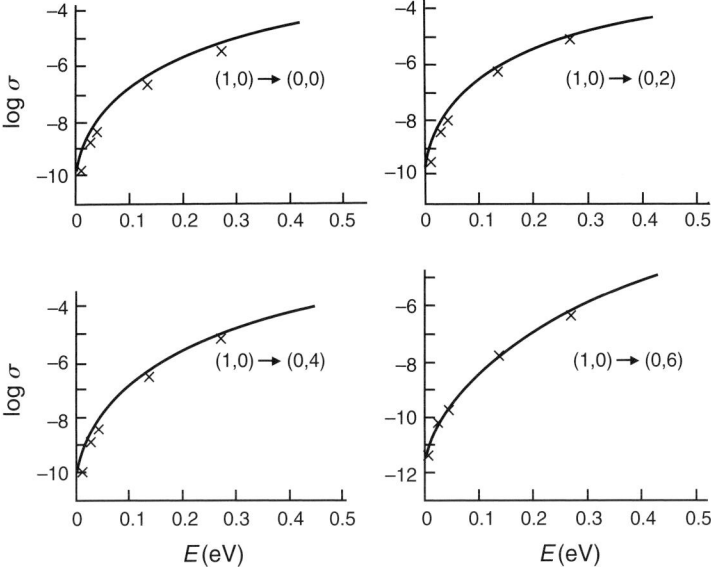

Figure 5.1 A comparison of CC(×) and CS calculations of rovibrational cross-sections for $(v,j) = (1,0) \rightarrow (0,j)$ transitions in He–H$_2$ collisions [74,75]. Cross-sections, σ, in units of 10^{-16} cm^2 and energies, E, in eV, relative to the $(1,0)$ level.

where $k_v^2 \equiv k_{v0}^2$ and $y_\lambda(v, v'|R) \equiv y_\lambda(v0, v'0|R)$. The equations (5.23) are coupled only through the vibrational quantum number, v. As shown in Chapter 4, their solution, for given values of θ' and l, yields an angle-dependent scattering matrix, **S**, with elements $S_l(v, v'|\theta')$. By analogy with equation (4.73), angular quadrature then yields

$$S_l(vj\Omega, v'j'\Omega') - \delta_{\Omega\Omega'} 2\pi \int_{\theta'=0}^{\pi} Y_{j\Omega}^*(\theta', 0) S_l(v, v'|\theta') \\ \times Y_{j'\Omega}(\theta', 0) \sin\theta' \, d\theta' \quad (5.24)$$

Elements of the transmission matrix, **T**, relate to elements of the scattering matrix, **S**, through

$$T_l(vj\Omega, v'j'\Omega') = \delta_{vv'}\delta_{jj'}\delta_{\Omega\Omega'} - S_l(vj\Omega, v'j'\Omega') \quad (5.25)$$

and the total cross-section is evaluated from

$$\sigma(vj \leftarrow v'j') = \frac{\pi}{k_{v'j'}^2 (2j'+1)} \sum_{l\Omega} (2l+1)|T_l(vj\Omega, v'j'\Omega)|^2 \quad (5.26)$$

A useful consequence of using the IOS approximation is the factorization of cross-sections for $vj \leftarrow v'j'$ in terms of those for which $j' = 0$:

$$\sigma(vj \leftarrow v'j') = \frac{k_{v'0}^2}{k_{v'j'}^2} \sum_{j''} \left(C_{000}^{j'j''j} \right)^2 \sigma(vj'' \leftarrow v'0) \quad (5.27)$$

88 The vibrational excitation of linear molecules

[76, 77]. Summing over j and using the orthonormality of the Clebsch–Gordan coefficients, $C_{000}^{j'j''j}$, we obtain

$$\sum_j \sigma(vj \leftarrow v'j') = \frac{k_{v'0}^2}{k_{v'j'}^2} \sum_j \sigma(vj \leftarrow v'0) \tag{5.28}$$

Equation (5.28) is an expression for the rotationally summed cross-section, i.e., the cross-section from an initial rovibrational state $v'j'$, to the final vibrational manifold, v. It should be recalled that this equation derives from the use of the IOS approximation.

In order to interpret experimental measurements or astronomical observations, the rate coefficient is usually required. Assuming only that the distribution of particle velocities is Maxwellian at kinetic temperature T, the rate coefficient, $\langle \sigma v \rangle$, is related to the cross-section, σ, by

$$\langle \sigma v \rangle = \int_0^\infty v\sigma(v) f(v,T) \, dv \tag{5.29}$$

where v denotes the collision velocity and

$$f(v,T) = 4\pi \left(\frac{\mu}{2\pi k_B T} \right)^{\frac{3}{2}} v^2 \exp\left(-\frac{\mu v^2}{2 k_B T} \right) \tag{5.30}$$

is the velocity distribution. Defining $x = \mu v^2/(2 k_B T)$, equation (5.29) may be written as

$$\langle \sigma v \rangle = \left(\frac{8 k_B T}{\pi \mu} \right)^{\frac{1}{2}} \int_0^\infty x \sigma(x) e^{-x} \, dx \tag{5.31}$$

where $[8 k_B T/(\pi \mu)]^{\frac{1}{2}}$ is the mean relative collision velocity; μ is the reduced mass. From equation (5.28), it may be seen that

$$\sum_j \sigma(vj \leftarrow v'j') x_{v'j'} = \sum_j \sigma(vj \leftarrow v'0) x_{v'0} \tag{5.32}$$

and hence, using (5.31)

$$\sum_j \langle \sigma v \rangle_{vj \leftarrow v'j'} = \sum_j \langle \sigma v \rangle_{vj \leftarrow v'0} \tag{5.33}$$

Equation (5.33) states that the rate coefficient for the vibrational transition $v \leftarrow v'$, summed over all the rotational states j in the final vibrational manifold, v, is independent of the initial rotational state, j'. This simplification is a direct consequence of the IOS approximation.

Under laboratory conditions, the gas density is often sufficiently high for a Boltzmann distribution of rotational level populations to apply. Then, the rate coefficient for a vibrational transition may be determined from

$$\langle \sigma v \rangle_{v \leftarrow v'} = Z^{-1} \sum_{jj'} (2j'+1) \exp\left(-\frac{\epsilon_{v'j'} - \epsilon_{v'0}}{k_B T} \right) \langle \sigma v \rangle_{vj \leftarrow v'j'} \tag{5.34}$$

5.2 The scattering of an atom by a vibrating rotor

where ϵ denotes a rovibrational eigenenergy and Z is the rotational partition function,

$$Z = \sum_{j'} (2j' + 1) \exp\left(-\frac{\epsilon_{v'j'} - \epsilon_{v'0}}{k_B T}\right) \quad (5.35)$$

Finally, we note that the principle of detailed balance implies that the rates of rovibrational excitation and de-excitation are related by

$$\langle \sigma v \rangle_{v'j' \leftarrow vj} (2j + 1) \exp\left(-\frac{\epsilon_{vj}}{k_B T}\right)$$
$$= \langle \sigma v \rangle_{vj \leftarrow v'j'} (2j' + 1) \exp\left(-\frac{\epsilon_{v'0}}{k_B T}\right) \quad (5.36)$$

5.2.2 The rovibrational excitation of H$_2$ by He

The H$_2$–He interaction potential has been the subject of a number of *ab initio* calculations over the years. In spite of the relative simplicity of this (four-electron) system, accurate results remain difficult to obtain; this is especially true of the region of the shallow potential minimum, which arises from the compensation of the long-range Van der Waals attraction by the short-range coulomb repulsion. Indeed, this difficulty is common to all neutral–neutral systems comprising a closed-shell target and a closed-shell perturber. The region of the potential minimum is particularly important in the context of the rotational excitation of heavy molecules (with small rotational constants) at low collision energies. Vibrationally inelastic scattering, or rotationally inelastic scattering of a light molecule, such as H$_2$, occurs at higher collision energies, at which the form of the interaction shortwards of the potential minimum assumes importance; the potential in this region can be calculated more readily.

Muchnick and Russek [78] performed *ab initio* calculations of the H$_2$–He potential and provided a parametric fit to their results, incorporating the dependence on the intramolecular (vibrational) coordinate, r. This potential was used subsequently [79] to compute cross-sections and hence rate coefficients for transitions, induced by He, between the rovibrational levels of H$_2$ up to approximately 20 000 K above the ground state, i.e. up to $(v, j) = (3, 7)$ of ortho-H$_2$ and $(v, j) = (3, 8)$ of para-H$_2$. The quantum mechanical CC method was employed, without further dynamical approximations. The vibrational eigenstates were represented by harmonic oscillator functions [80]. Spectroscopically measured values of the energies of the rovibrational levels [81] were used. The rate coefficients for rotational transitions within the $v = 0$ vibrational ground state were compared with the calculations of Schaefer and Köhler [82] and found to agree to within a few per cent.

The rate coefficient for vibrational relaxation, $v = 1 \to 0$, summed over the final rotational states and averaged over the initial rotational states is

$$q_{v=0 \leftarrow v'=1}(T) = \frac{\sum_{j,j'} q(v = 0, j \leftarrow v' = 1, j') n(v' = 1, j')}{\sum_{j'} n(v' = 1, j')} \quad (5.37)$$

and assuming a Boltzmann distribution of the level populations, $n(v' = 1, j')$, within the $v' = 1$ manifold, was also calculated in the kinetic temperature range $100 < T < 3000$ K. The results were found to agree with laboratory measurements [83, 84] to within a factor of 2 over this

The vibrational excitation of linear molecules

Table 5.1. *Computed values of the rate coefficients (in units of $cm^3\ s^{-1}$) for vibrational relaxation of $H_2(v=1)$ in collisions with He atoms. For each value of the temperature, the upper entry is for ortho-H_2 and the lower entry is for para-H_2. Numbers in parentheses are powers of 10.*

T (K)	$q_{v=0 \leftarrow v'=1}$
100	1.3(−18)
	7.4(−19)
300	4.5(−17)
	5.1(−17)
500	5.1(−16)
	5.2(−16)
1000	1.2(−14)
	1.2(−14)
3000	1.1(−12)
	1.1(−12)

entire range of temperatures. The computed values of the rate coefficients, for transitions in ortho- and para-H_2, are given in Table 5.1. The rate coefficients increase by six orders of magnitude between $T = 100$ K and $T = 3000$ K.

Independent quantum mechanical calculations have been performed by Balakrishnan *et al.* [85, 86], using numerically exact vibrational eigenfunctions and extending up to $v = 6$; their results are in generally good agreement with [79]. It may be concluded that the rate coefficients for pure rotational transitions in H_2, induced by He, are known to high accuracy, and those for rovibrational transitions to an accuracy which is acceptable for many astrophysical applications, particularly for the purpose of determining the contribution to the H_2 cooling function (see Section 5.4 below).

5.2.3 The rovibrational excitation of H_2 by H

Although it is ostensibly the simplest atom–molecule system, the scattering of H on H_2 proves to be difficult to treat theoretically. At low energies, the cross-sections for pure rotational transitions are small, owing to the weak anisotropy (θ'-dependence) of the interaction potential. These cross-sections are found also to be sensitive to the representation of the $v = 0$ vibrational eigenfunction [87, 88].

A further complication, at high collision energies, is that reactive scattering can occur. The reaction barrier is approximately 5000 K, and reactive scattering assumes importance at temperatures $T > 1000$ K. Calculations which incorporated this process have been performed, using either a quantum mechanical approach, but including only the rotational levels $j \leq 3$

5.2 The scattering of an atom by a vibrating rotor

of $v = 0$ [89], or the quasi-classical trajectory (QCT) Monte Carlo method [90–92]. The QCT method combines the use of classical mechanics, to treat the scattering process, with Monte Carlo sampling of the initial conditions. Quantization is simulated by means of a 'binning' procedure, which involves allocating the the final state to discrete values of the corresponding quantum numbers. However, the QCT method is only as valid as the classical mechanics that underpins it. As the collision energy falls, the cross-sections may fail to satisfy detailed balance, which is a symptom of the break-down of the method. On the other hand, the QCT approach is both appropriate and better adapted (than quantum mechanics) to obtaining results for large values of the quantum numbers, as anticipated from the correspondence principle.

At temperatures $T < 3000$ K, rate coefficients for vibrational relaxation in *non-reactive scattering*, determined by means of the quantum mechanical method [93], fall increasingly below the values that derive from the QCT method, perhaps indicating the progressive failure of the quasi-classical approach. At high energies, the reactive scattering channels should be included, and quantum mechanical calculations have become feasible only recently.

5.2.4 The rovibrational excitation of HD by He and H

The interaction potentials for H_2–He and H_2–H may be used to study HD–He and HD–H scattering, as the coulomb interactions with HD are the same as with H_2. Allowance must be made for the displacement of the centre of mass of HD molecule from the mid-point of its internuclear axis and for the effects of the higher value of the reduced molecular mass on the rovibrational eigenfunctions and eigenenergies.

The shift of the centre of mass has the important consequence that transitions involving odd values of Δj are allowed in *non-reactive* scattering on HD; this is consistent with the the fact that the nuclei (H, a fermion, and D, a boson) are distinguishable, and hence the ortho/para dichotomy that exists in the case of H_2 is absent in HD. Indeed, $\Delta j = 1$ transitions dominate rotational population transfer in HD. However, the absence of the ortho and para modifications implies that *all* the rotational states have to be included simultaneously in calculations of scattering on HD, which results in a considerable increase in computing time.

Schaefer [94] calculated rate coefficients for the pure rotational excitation of HD by He, but only for the rotational levels $j \leq 4$. Subsequently, Roueff and Zeippen [95] recomputed these data, using a more recent interaction potential and including the levels $v = 0, j \leq 9$. Cross-sections and rate coefficients for HD–H scattering, with the basis $v = 0, j \leq 9$, have also been calculated [96]. To the extent that comparison is possible, results for pure rotational transitions in HD–He scattering were found to agree well with the previous calculations of Schaefer [94]. For HD–H, the rate coefficients for transitions $\Delta j = 1$ and $\Delta j = 2$ were found to be similar in magnitude to those for HD–He scattering.

It has already been mentioned that transitions $\Delta j = 1$ dominate population transfer in HD. At $T = 100$ K, the rate coefficients for de-excitation by H of the first rotationally excited state of HD, $v = 0, j = 1$, to the ground state, $v = 0, j = 0$, is of the order of 10^{-11} cm^3 s^{-1}; this may be compared with values in the range 10^{-13} to 10^{-14} cm^3 s^{-1} for the transitions, induced by H, from the first rotationally excited states of ortho- and para-H_2 to their respective ground states, i.e. $v = 0, j = 3 \rightarrow 1$ and $v = 0, j = 2 \rightarrow 0$.

Flower and Roueff [97] extended the collision calculations for the systems HD–H to excited vibrational states of HD, including levels $v \leq 2, j \leq 9$. Roueff and Zeippen [98] studied

HD–He, including levels of HD up to $(v,j) = (3,3)$. Thus, the data relating to the collisional excitation of HD have a similar degree of completeness to those for H_2.

5.3 Excitation of H_2 and HD in collisions with H_2 molecules

In the molecular clouds of the interstellar medium, H_2 molecules are the most abundant species; their number density exceeds, by about a factor 5, the next most abundant species, helium. When the temperature of the gas is raised, for example, by the passage of a shock wave, rotationally inelastic collisions between H_2 molecules usually determine the rate of cooling of the hot gas. If the degree of dissociation of molecular hydrogen remains low, collisions between H_2 molecules dominate the process of vibrational excitation also.

Pure rotational excitation of H_2 by H_2 has been studied by a number of authors, including Schaefer and Meyer [99], Monchick and Schaefer [100] and Danby et al. [101]; these authors built on earlier work by Sheldon Green and his collaborators, taking advantage of more recent and more accurate determinations of the H_2–H_2 interaction potential. Subsequently, the results of Schwenke's fit to his own *ab initio* determination of the potential [102] were used to compute cross-sections and rate coefficients for pure rotational and for rovibrational transitions in H_2, induced by other H_2 molecules [103–105].

Transitions between the two forms of molecular hydrogen – ortho (j odd) and para (j even) – involve a change in the total nuclear spin ($I = 1$ in ortho, $I = 0$ in para) and are induced by proton-exchanging reactions of H_2 molecules with the ions H^+ and H_3^+ or with H atoms. In H_2–H_2 collisions, only non-reactive scattering needs to be considered, and the ortho and para forms remain distinct. It follows that ortho–ortho, ortho–para and para–para scattering can be considered separately.

In general, simultaneous vibrational excitation of both molecules in a collision is much less likely than single excitation. Accordingly, calculations may be performed of the rovibrational excitation of ortho-H_2 by ortho-H_2 and para-H_2, constrained to their vibrational ground state, $v = 0$, and of the rovibrational excitation of para-H_2 by ortho-H_2 and para-H_2, in their vibrational ground state. The coupled differential equations that have to be solved in this case are a generalization of equation (5.17) and may be written in the form

$$\left[\frac{d^2}{dR^2} - \frac{l(l+1)}{R^2} + k_{v_2 j_2 j_1}^2\right] F(v_2 j_2 j_1 j_{12} l p J | R)$$
$$= 2\mu \sum_{v_2' j_2' j_1' j_{12}' l'} V(v_2 j_2 j_1 j_{12} l, v_2' j_2' j_1' j_{12}' l'; pJ | R) F(v_2' j_2' j_1' j_{12}' l' pJ | R) \quad (5.38)$$

where

$$V(v_2 j_2 j_1 j_{12} l, v_2' j_2' j_1' j_{12}' l'; pJ | R)$$
$$= \sum_{\lambda_1 \lambda_2 \mu \geq 0} c_{\lambda_1 \lambda_2 \mu}(j_2 j_1 j_{12} l, j_2' j_1' j_{12}' l'; pJ) y_{\lambda_1 \lambda_2 \mu}(v_2 j_2, v_2' j_2' | R) \quad (5.39)$$

and

$$y_{\lambda_1 \lambda_2 \mu}(v_2 j_2, v_2' j_2' | R) = \int_0^\infty \chi^*(v_2 j_2 | r_2) v_{\lambda_1 \lambda_2 \mu}(r_2, R) \chi(v_2' j_2' | r_2) dr_2 \quad (5.40)$$

5.4 Cooling functions

determines the coupling between the rovibrational states $v_2 j_2$ and $v'_2 j'_2$; $k^2_{v_2 j_2 j_1} = 2\mu (E - \epsilon_{v_2 j_2} - \epsilon_{0 j_1})$, where the ϵ_{vj} are the eigenenergies of the H$_2$ rovibrational states. The algebraic coefficients $c_{\lambda_1 \lambda_2 \mu}(j_2 j_1 j_{12} l, j'_2 j'_1 j'_{12} l'; pJ)$ are generalizations of the Percival–Seaton coefficients, which have already been encountered [equation (4.56)]; they are expressible in terms of Wigner 3j-, 6j- and 9j-coefficients [see chapter 36 of [43]]. It is assumed that the H$_2$–H$_2$ interaction potential has been expanded in terms of spherical harmonic functions of BF coordinates, in the form

$$V(\hat{\mathbf{r}}'_1, \mathbf{r}'_2, R) = \sum_{\lambda_1 \lambda_2 \mu \geq 0} v_{\lambda_1 \lambda_2 \mu}(r_2, R)$$

$$\times \frac{4\pi}{[2(1+\delta_{\mu 0})]^{\frac{1}{2}}} [Y_{\lambda_1 \mu}(\hat{\mathbf{r}}'_1) Y_{\lambda_2, -\mu}(\hat{\mathbf{r}}'_2) + Y_{\lambda_1, -\mu}(\hat{\mathbf{r}}'_1) Y_{\lambda_2 \mu}(\hat{\mathbf{r}}'_2)] \quad (5.41)$$

Equations (5.38) are to be solved for each value of the total angular momentum of the system

$$\mathbf{J} = \mathbf{j}_{12} + \mathbf{l}$$

where

$$\mathbf{j}_{12} = \mathbf{j}_1 + \mathbf{j}_2$$

and for each value of the total parity, $p = (-1)^{j_1 + j_2 + l}$.

For pure rotational transitions within the $v = 0$ ground vibrational manifold, the more recent determinations of the rate coefficients agree with the earlier calculations [100, 101] to better than 30%. Regarding vibrational relaxation, the level of agreement with the measurements [84, 106, 107] is satisfactory at low temperatures ($T \approx 100$ K) and at high temperatures ($T > 1000$ K), but, at $T = 500$ K, the calculated value is almost an order of magnitude smaller than measured. This discrepancy might be attributable to inaccuracies in the H$_2$–H$_2$ interaction potential or in the now 30-year-old measurements.

The interaction potential for H$_2$–H$_2$ has been used [108] to study HD–H$_2$ scattering, by making the appropriate allowances for the isotopic substitution of D for H. Levels of HD with $v \leq 2, J \leq 9$ were included in these calculations. These data are necessary to evaluate the rate of cooling by HD reliably, notably in cold molecular gas where chemical fractionation in favour of HD has occurred (see sub-section entitled 'Dense clouds' in Section 1.2.2).

5.4 Cooling functions

Molecular hydrogen is an important coolant of molecular gas that has been heated to temperatures of a few hundred degrees or more by events such as the passage of a shock wave. Even when partially dissociated, at temperatures of the order of 1000 K, H$_2$ may still dominate the cooling, owing to the high intrinsic elemental abundance of hydrogen. However, as the kinetic temperature falls towards 100 K, the wide spacing of the rotational levels of H$_2$ begins to counter its high abundance, and heavier molecules, such as HD, H$_2$O and CO, play increasingly important roles in the thermal balance of the gas.

Three effects intervene when comparing the rates of cooling of gas by H$_2$ and HD at temperatures T of the order of 100 K or less.

- Owing to its higher reduced mass, the rotational constant of HD is smaller, by a factor of approximately 4/3, than that of H_2.
- Owing to the fact that it is not homonuclear, $\Delta j = 1$ transitions can be excited collisionally in HD, whereas they occur in H_2 only in reactive scattering events. Thus, the energy of the lowest rotational transition of HD ($j = 0 \rightarrow 1$) is approximately a factor of 4 less than the energy of the lowest transition ($j = 0 \rightarrow 2$) in H_2.
- Chemical fractionation can occur at low temperatures, essentially because of the difference in the zero-point vibrational energies, $\hbar\omega/2$, of HD and H_2: the vibrational frequency, $\omega \propto (1/m)^{\frac{1}{2}}$, is smaller by a factor $(4/3)^{\frac{1}{2}} \approx 1.15$ in HD than in H_2. This difference in the zero-point energies translates into an exothermicity of approximately 400 K of reactions in which the isotopic substitution of deuterium for hydrogen occurs – and an *endo*thermicity of the same magnitude for the reverse reactions. At temperatures of the order of 100 K, the abundance of HD can become enhanced, relative to that of H_2, so that $n(\text{HD})/n(H_2) \gg n_\text{D}/n_\text{H}$. However, we note that this effect, of chemical fractionation, is important only in *partially* molecular gas. In regions where both H and D are essentially in molecular form, $n(\text{HD})/n(H_2) \approx 2n_\text{D}/n_\text{H}$.

5.4.1 Results

For the purpose of applications, the rate coefficients, $\langle \sigma v \rangle$, for collisionally induced transitions are often fitted to simple functions of the kinetic temperature, T. In the cases of H_2 and HD, a form that has been found to be suitable is

$$\log \langle \sigma v \rangle = a + b/t + c/t^2$$

where $t = 10^{-3}T + \delta t$ and a, b, c are transition-dependent constants; δt, which is also a constant, is introduced in order to prevent divergence of the fit at low temperatures. δt depends on the collision system (e.g. HD–H) but is independent of the transition.

Knowing the rate coefficients, spontaneous radiative transition probabilities, and the spectroscopic values of the energy levels, the level populations (of H_2 and HD) may be computed, in steady state, for given values of the total density, $n_\text{H} = n(\text{H}) + 2n(H_2) + n(\text{HD}) + n(H^+)$, kinetic temperature, T, atomic to molecular hydrogen density ratio, $n(\text{H})/n(H_2)$, and ortho to para density ratio, $n(\text{ortho-}H_2)/n(\text{para-}H_2)$. The rate of cooling per H_2 or HD molecule, the *cooling function*, is then given by

$$W(X) = \frac{1}{n(X)} \sum_{i>j} (E_i - E_j) n_i A(i \rightarrow j)$$

where $X = H_2$ or HD, E_i is the energy of level i, relative to the ground state, n_i is the density of population in this level, and $A(i \rightarrow j)$ is the spontaneous radiative transition probability from level i to a lower level j; $W(X)$ is usually expressed in units of erg s^{-1} (10^{-7} W). In this approach, it is implicitly assumed that the opacities of the transitions, and hence self-absorption, may be neglected. This assumption is often valid: the rovibrational spectrum of H_2 is quadrupole, as H_2 has no permanent dipole moment; HD, which has a lower abundance

5.4 Cooling functions

than H_2, has only a weak dipole moment. For an electric dipole rovibrational transition within a Σ electronic state (the ground electronic state of HD), the change in the rotational quantum number is limited to $\Delta j = 0, \pm 1$ (but $\Delta j = 0$, when $j = 0$, is not allowed). For a transition from rovibrational state $v', j+1$ to v, j, i.e. a $v' - v$ R(j) transition, the radiative transition probability is given by

$$A(v,j \leftarrow v',j+1) = \frac{4(E_{v',j+1} - E_{v,j})^3}{3\hbar^4 c^3} \frac{j+1}{2j+3} D^2_{v',j+1;v,j} \tag{5.42}$$

where $D_{v',j+1;v,j}$ is the matrix element of the dipole moment (see Section 10.3). For a $v' - v$ Q(j) transition, that is from v', j to v, j, the corresponding expression for the transition probability is

$$A(v,j \leftarrow v',j) = \frac{4(E_{v',j} - E_{v,j})^3}{3\hbar^4 c^3} \frac{1}{j(j+1)} D^2_{v',j;v,j} \tag{5.43}$$

Q(j) transitions are not possible within a given vibrational manifold, i.e. when $v' = v$. Finally, for a $v' - v$ P(j) transition, from $v', j-1$ to v, j, the transition probability is

$$A(v,j \leftarrow v',j-1) = \frac{4(E_{v',j-1} - E_{v,j})^3}{3\hbar^4 c^3} \frac{j}{2j-1} D^2_{v',j-1;v,j} \tag{5.44}$$

Equation (5.42), for example, may be written as

$$A(v,j \leftarrow v',j+1) = 2.14 \times 10^{10} (E_{v',j+1} - E_{v,j})^3 \frac{j+1}{2j+3} D^2_{v',j+1;v,j}$$

where the energies E and the dipole moment D are expressed in atomic units (the atomic unit of energy is the hartree, and of the dipole moment is 2.54 debye); A is in s^{-1}. For electric quadrupole transitions, the change in the rotational quantum number is $\Delta j = 0, \pm 2$ (but $\Delta j = 0$, when $j = 0$, is not allowed). The transition probabilities, which are proportional to the square of the matrix elements of the quadrupole moment (see Section 10.3), are given by

$$A(v,j \leftarrow v',j+2) = \frac{(E_{v',j+2} - E_{v,j})^5}{15\hbar^6 c^5} \frac{3(j+2)(j+1)}{2(2j+5)(2j+3)} Q^2_{v',j+2;v,j} \tag{5.45}$$

$$A(v,j \leftarrow v',j) = \frac{(E_{v',j} - E_{v,j})^5}{15\hbar^6 c^5} \frac{j(j+1)}{(2j+3)(2j-1)} Q^2_{v',j;v,j} \tag{5.46}$$

$$A(v,j \leftarrow v',j-2) = \frac{(E_{v',j-2} - E_{v,j})^5}{15\hbar^6 c^5} \frac{3j(j-1)}{2(2j-1)(2j-3)} Q^2_{v',j-2;v,j} \tag{5.47}$$

for the $v' - v$ S(j), $v' - v$ Q(j) and $v' - v$ O(j) transitions, respectively. If the energies E and the quadrupole moment Q are expressed in atomic units, then equation (5.45), for example, becomes

$$A(v,j \leftarrow v',j+2) = 1.43 \times 10^4 (E_{v',j+2} - E_{v,j})^5 \frac{3(j+2)(j+1)}{2(2j+5)(2j+3)} Q^2_{v',j+2;v,j}$$

where A is in s^{-1}. We recall that the dipole moment, of H_2 is zero, whereas that of HD is small but finite; for transitions within the vibrational ground state of HD, the dipole moment matrix element is approximately equal to 3.4×10^{-4}, in atomic units [109]. For transitions within the vibrational ground state of H_2, the quadrupole moment matrix element is approximately 0.97, in atomic units [110]. Accurate calculations require the Schrödinger equation for the rovibrational motion of the molecule [equation (5.8)] to be solved numerically; the dipole and quadrupole matrix elements vary with j, owing to the presence of the centrifugal potential.

In Figs 5.2 and 5.3, we plot the cooling functions $W(H_2)$ and $W(HD)$, respectively, as functions of T, for $1 \leq n_H \leq 10^8$ cm^{-3}, and specific values of $n(\text{ortho-}H_2)/n(\text{para-}H_2) = 1$, and $n(H)/n(H_2) = 1$. Particularly at low temperatures and high densities, the cooling rate per HD molecule is much larger than the cooling rate per H_2 molecule. The smaller rotational constant (and hence closer rotational level spacing) and larger radiative transition probabilities

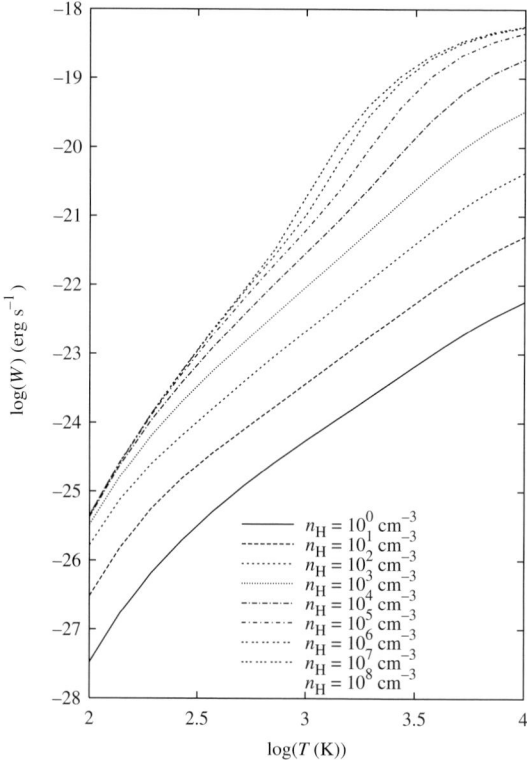

Figure 5.2 The cooling function, $W(H_2)$ (in units of 10^{-7} W), calculated for $1 \leq n_H \leq 10^8$ cm^{-3}, an ortho : para H_2 density ratio of 1, and a H/H_2 abundance ratio of 1 [111].

5.4 Cooling functions

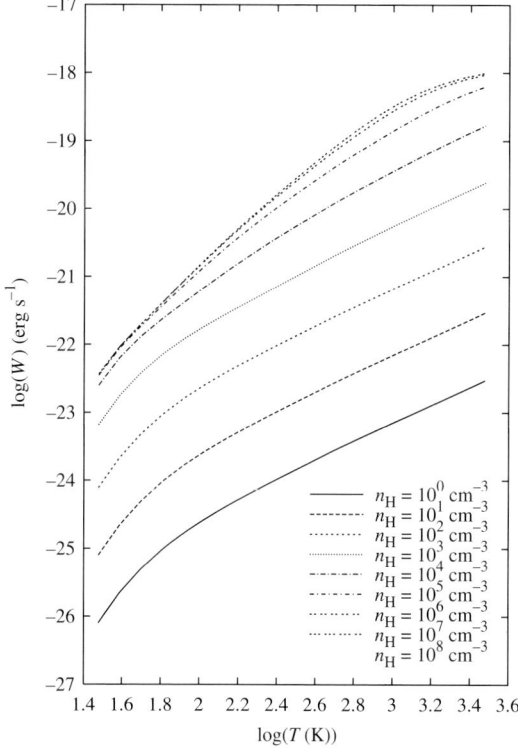

Figure 5.3 As in Fig. 5.2, but for the cooling function $W(\mathrm{HD})$, in units of 10^{-7} W [112].

ensure that, as the Boltzmann limit is approached at low temperatures, a molecule of HD is a more efficient coolant than a molecule of H_2. Of course, the rate of cooling *per unit volume* is proportional to the molecular number density, $n(H_2)$ or $n(\mathrm{HD})$. In media in which the transformation of H into H_2, and D into HD, is incomplete, chemical fractionation can, at low temperatures, enhance the ratio $n(\mathrm{HD})/n(H_2)$ relative to the elemental abundance ratio, $n_D/n_H \approx 10^{-5}$. In the primordial gas, for example, chemical fractionation may have increased the ratio of the densities of HD and H_2 to $n(\mathrm{HD})/n(H_2) \approx 10^{-3}$ [113]. Under these circumstances, the contribution of HD to the cooling of the medium cannot be neglected.

6

The excitation of fine structure transitions

6.1 Introduction

An important cooling process in interstellar clouds is the excitation of fine structure transitions in abundant atoms and ions, followed by radiative decay. The relevant transitions are those between the fine structure components of ground terms, such as C^0 $2p^2$ ^3P, O^0 $2p^4$ ^3P and C^+ $2p$ $^2P^o$. By *term* is meant the *LS*-coupling state denoted by ^{2S+1}L, where L is the total electronic orbital angular momentum quantum number and S is the total electronic spin angular momentum quantum number. Only the outer (valence) electrons (e.g. $2p^2$) need to be listed, as the inner shells and sub-shells are closed and have zero resultant angular momenta. Spectrosopic notation is used to denote the orbital angular momentum: 's' for $l = 0$, 'p' for $l = 1$, 'd' for $l = 2$, ..., with upper case letters indicating resultant angular momenta (vector sums of the contributions of the individual valence electrons). Departures from *LS*-coupling, owing to the *spin–orbit interaction*, result in the states with different values of J, the total electronic angular momentum (the vector sum of the orbital and spin angular momenta) having slightly different energies; these *fine structure* states are denoted $^{2S+1}L_J$. Thus the ground ^3P term of C^0 and O^0 is a triplet, comprising the three fine structure components with $J = 0, 1, 2$, and the $^2P^o$ ground term of C^+ is a doublet with $J = 1/2, 3/2$. (For complementary information, see Chapter 9.)

The parity of the states, $\sum_i l_i$, is denoted by the superscript 'o', when it is odd, and is usually omitted (and implied) when even. All fine structure components of a given term have the same parity, and so electric dipole transitions, which require that the initial and final states have different parity, are forbidden between the components of a given term. On the other hand, magnetic dipole or electric quadrupole transitions are allowed, although with lower transition probabilities than would apply to the case of electric dipole radiation. For complementary information, see Sections 9.2 and 10.3.

In Table 6.1 the wavelengths and radiative transition probabilities of the transitions that are known to be important in the thermal balance of the interstellar gas are listed. The distinction between the notation of the atom or ion and the spectroscopic designation of its spectrum is made in this table. In much of the astronomy literature, this distinction is ignored. These fine structure transitions may be excited in collisions with electrons or heavy particles, principally H, H_2 and He. Electron collisional excitation will be discussed in Chapter 9. In the present chapter, we consider heavy particle collisions.

The main source of uncertainty in calculations of the cross-sections for the excitation of fine structure transitions by heavy particle impact is the interaction potential. This process is essentially *diabatic* (non-adiabatic) in nature, as it implies a change in the electronic state of the atom or ion. As a consequence, at least two adiabatic potential energy curves (or surfaces)

6.2 Theory of fine structure excitation processes

Table 6.1. *Atomic data relating to fine structure transitions in abundant atoms and ions. The spectroscopic notation for the spectrum of the atom or ion is given.* λ *denotes the wavelength of the transition,* Δ *the energy difference between the levels involved in the transition, and A the transition probability.* C^0*: [114];* C^+*: [115];* O^0*: [116];* Si^+*: [117];* Fe^+*: [118]. Numbers in parentheses are powers of 10.*

Atom/ion	Spectrum	Transition	$\lambda(\mu m)$	Δ (K)	A (s^{-1})
C^0	C I	$^3P_1 \to {}^3P_0$	610	24	7.93 (−8)
		$^3P_2 \to {}^3P_1$	370	39	2.65 (−7)
C^+	C II	$^2P^o_{3/2} \to {}^2P^o_{1/2}$	158	91	2.29 (−6)
O^0	O I	$^3P_1 \to {}^3P_2$	63.1	228	8.95 (−5)
		$^3P_0 \to {}^3P_1$	147	98	1.70 (−5)
Si^+	Si II	$^2P_{3/2} \to {}^2P_{1/2}$	34.8	413	2.17 (−4)
Fe^+	Fe II	$^6D_{7/2} \to {}^6D_{9/2}$	26.0	554	2.13 (−3)
		$^6D_{5/2} \to {}^6D_{7/2}$	35.3	407	1.57 (−3)

are involved. Thus, a knowledge of the potential energy curves for the ground state and perhaps several excited states of the collision complex is a pre-requisite for calculations of this type.

6.2 Theory of fine structure excitation processes

6.2.1 Systems with one open shell

We consider first the excitation of fine structure transitions in an atom or ion comprising one open shell, induced in collisions with closed-shell perturbers. Examples of such systems are Mg(3s3p $^3P^o$) + He, O(2p^4 ^3P) + He, and C$^+$(2p $^2P^o$) + H$_2$. A quantum mechanical theory appropriate to treating collisional processes in this category was given by Launay [119]. This theory is applicable not only to problems of fine structure excitation, but also to rotational excitation in collisions between molecules and to the study of molecular dimers; the corresponding FORTRAN code is MOLCOL [55].

Consider two systems of arbitrary angular momenta, $\mathbf{j}_1$ and $\mathbf{j}_2$. The isolated systems satisfy a Schrödinger equation of the type

$$h_i |\alpha_i j_i m_i\rangle = \epsilon_i |\alpha_i j_i m_i\rangle \qquad (i = 1, 2) \tag{6.1}$$

where h_i is the internal hamiltonian operator, m_i is the projection of $\mathbf{j}_i$ on the space-fixed (SF) z-axis, and α_i denotes all other quantum numbers associated with the remaining degrees of freedom of the system i; ϵ_i is the eigenenergy.

The internal angular momenta, $\mathbf{j}_i$, may be vector coupled to yield a resultant

$$\mathbf{j}_{12} = \mathbf{j}_1 + \mathbf{j}_2 \tag{6.2}$$

The corresponding wave function is

$$|\gamma j_{12} m_{12}\rangle = \sum_{m_1 m_2} C^{j_1 j_2 j_{12}}_{m_1 m_2 m_{12}} |\alpha_1 j_1 m_1\rangle |\alpha_2 j_2 m_2\rangle \tag{6.3}$$

where $C^{j_1 j_2 j_{12}}_{m_1 m_2 m_{12}}$ is a Clebsch–Gordan coefficient, and γ denotes $\alpha_1 j_1 \alpha_2 j_2$. The total angular momentum of the system is

$$\mathbf{J} = \mathbf{j}_{12} + \mathbf{l} \tag{6.4}$$

where $\mathbf{l}$ is the relative orbital angular momentum of 1 and 2. It follows that the wave function may be written

$$|\gamma j_{12} l J M\rangle = \sum_{m_{12} m_l} C^{j_{12} l J}_{m_{12} m_l M} |\gamma j_{12} m_{12}\rangle Y_{l m_l}(\Theta, \Phi) \tag{6.5}$$

where m_l and M are the projections of $\mathbf{l}$ and $\mathbf{J}$, respectively, on the SF z-axis, Y denotes a normalized spherical harmonic [cf. equation (4.10)], and (R, Θ, Φ) are the spherical polar coordinates of the intermolecular vector, $\mathbf{R}$, relative to the SF coordinate frame. The eigenfunction (6.5) has parity $p = (-1)^l p_1 p_2$, where p_i is the parity of system i.

In the discussion of atom–diatom scattering in Chapter 4, the problem was formulated in both the SF and the body-fixed (BF) coordinate systems. The BF Z-axis is taken to coincide with the intermolecular vector $\mathbf{R}$, i.e., with the vector joining the centres of mass of the two systems 1 and 2 (see Fig. 6.1).

In terms of BF coordinates, the wave function has the form

$$|\gamma j_{12} \Omega J M\rangle = \left(\frac{2J+1}{4\pi}\right)^{\frac{1}{2}} D^{J*}_{M\Omega}(\Phi, \Theta, 0) |\gamma j_{12} \Omega\rangle \tag{6.6}$$

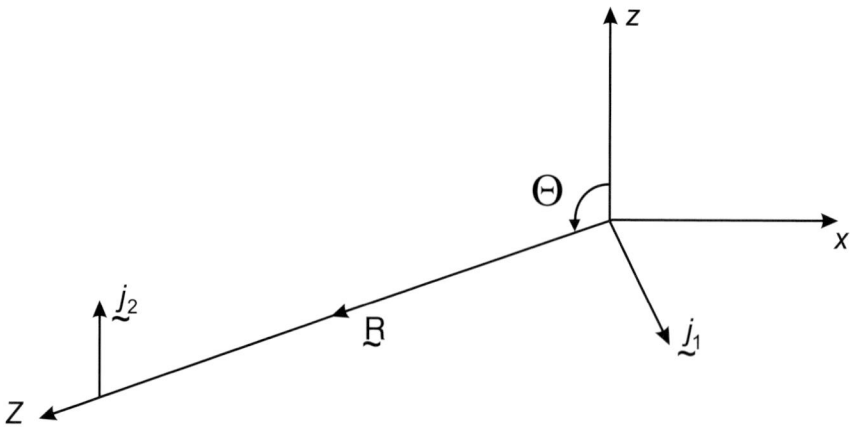

Figure 6.1 The interaction between two systems of arbitrary angular momenta, $\mathbf{j}_1$ and $\mathbf{j}_2$. The body-fixed Z-axis is taken to lie along the vector $\mathbf{R}$ which joins the centre of mass of system 1 to the centre of mass of system 2. The polar angles of $\mathbf{R}$ are (Θ, Φ) relative to the space-fixed (xyz) coordinate frame.

6.2 Theory of fine structure excitation processes

where $|\gamma j_{12} \Omega\rangle$ is a function of BF coordinates, Ω is the projection of $\mathbf{j}_{12}$ on the BF Z-axis, and **D** is the rotation matrix [48,49] (see Chapter 4). Under the action of the inversion operator, P, the BF wave function (6.6) transforms as [119]

$$P|\gamma j_{12}\Omega JM\rangle = (-1)^{J-j_{12}} p_1 p_2 |\gamma j_{12}, -\Omega JM\rangle \tag{6.7}$$

Thus, equation (6.6) is not an eigenfunction of P, but the linear combinations

$$|\gamma j_{12}\bar{\Omega}\epsilon JM\rangle = \frac{|\gamma j_{12}\bar{\Omega}JM\rangle + \epsilon|\gamma j_{12}, -\bar{\Omega}JM\rangle}{[2(1+\delta_{\bar{\Omega}0})]^{\frac{1}{2}}} \tag{6.8}$$

are eigenfunctions of P with eigenvalues $p' = (-1)^{J-j_{12}} \epsilon p_1 p_2$. In equation (6.8), $\bar{\Omega} = |\Omega|$ and $\epsilon = \pm 1$; $[2(1+\delta_{\bar{\Omega}0})]^{-\frac{1}{2}}$ is a normalization factor. When $\Omega = 0$, only $\epsilon = +1$ is allowed.

The SF functions (6.5) and the BF functions (6.8) are related through a unitary transformation. The elements of the transformation matrix are given by

$$\langle \gamma j_{12}\bar{\Omega}\epsilon JM | \gamma j_{12} lJM\rangle = \left[\frac{(2l+1)}{2(1+\delta_{\bar{\Omega}0})(2J+1)}\right]^{\frac{1}{2}} C^{j_{12}lJ}_{\bar{\Omega}0\bar{\Omega}}(1+pp') \tag{6.9}$$

and they vanish when $p \neq p'$ (i.e. when $pp' = -1$). Put another way, because the parity of the wave function is invariant under the rotation of the coordinate system that takes the SF into the BF frame, we must have that $p' = p$, in which case $\epsilon = [(-1)^{J-j_{12}} p_1 p_2] p$, and the BF wave function (6.8) may be written in terms of the total parity, p:

$$|\gamma j_{12}\bar{\Omega} pJM\rangle = |\gamma j_{12}\bar{\Omega}, [(-1)^{J-j_{12}} p_1 p_2] p, JM\rangle \tag{6.10}$$

This form of the wave function will be used in the following analysis.

By analogy with equation (4.36), the total wave function, Ψ, may be expanded in terms of the basis functions (6.10)

$$\Psi = \sum_{\gamma j_{12}\bar{\Omega} pJM} \frac{G(\gamma j_{12}\bar{\Omega} pJM | R)}{R} |\gamma j_{12}\bar{\Omega} pJM\rangle \tag{6.11}$$

and is required to satisfy the Schrödinger equation for the scattering system

$$H\Psi = E\Psi \tag{6.12}$$

where

$$H = h_1 + h_2 - \frac{\nabla_R^2}{2\mu} + V \tag{6.13}$$

and E is the total energy in the centre of mass system. In (6.13), μ is the reduced mass of the collision complex of systems 1 and 2, R is the separation of the centres of mass of 1 and 2, and V is the interaction potential, to be considered below.

The excitation of fine structure transitions

Using procedures introduced in Chapter 4, the Schrödinger equation (6.12) may be reduced to a set of coupled second-order differential equations:

$$\left[\frac{d^2}{dR^2} + k_\gamma^2\right] G(\gamma j_{12}\bar{\Omega}pJM|R)$$

$$= 2\mu \sum_{\gamma' j'_{12}\bar{\Omega}'p'J'M'} \langle \gamma j_{12}\bar{\Omega}pJM | V + \frac{\mathbf{l}^2}{2\mu R^2} | \gamma' j'_{12}\bar{\Omega}'p'J'M' \rangle$$

$$\times G(\gamma' j'_{12}\bar{\Omega}'p'J'M'|R) \tag{6.14}$$

The non-vanishing elements of the angular momentum operator, $\mathbf{l}^2$, may be derived from equations (4.63) and (4.64), generalized to the case of half-integral angular momenta:

$$\langle \gamma j_{12}\bar{\Omega}pJM | \mathbf{l}^2 | \gamma j_{12}\bar{\Omega}pJM \rangle = [J(J+1) + j_{12}(j_{12}+1) - 2\bar{\Omega}^2]$$

$$- \epsilon \delta_{\bar{\Omega},\frac{1}{2}} [J(J+1) - \bar{\Omega}(\bar{\Omega}-1)]^{\frac{1}{2}} [j_{12}(j_{12}+1) - \bar{\Omega}(\bar{\Omega}-1)]^{\frac{1}{2}} \tag{6.15}$$

and

$$\langle \gamma j_{12}\bar{\Omega}pJM | \mathbf{l}^2 | \gamma j_{12}, \bar{\Omega} \pm 1, pJM \rangle$$

$$= -(1 + \epsilon \delta_{\bar{\Omega}0})^{\frac{1}{2}} (1 + \epsilon \delta_{\bar{\Omega}\pm 1,0})^{\frac{1}{2}} [J(J+1) - \bar{\Omega}(\bar{\Omega}\pm 1)]^{\frac{1}{2}}$$

$$\times [j_{12}(j_{12}+1) - \bar{\Omega}(\bar{\Omega}\pm 1)]^{\frac{1}{2}}$$

$$+ \epsilon \delta_{\bar{\Omega},\frac{1}{2}} [J(J+1) + j_{12}(j_{12}+1) - 2\bar{\Omega}^2] \tag{6.16}$$

where $\epsilon = [(-1)^{J-j_{12}} p_1 p_2] p$, as noted above. In addition, the matrix elements of the potential, V, must be evaluated, and we turn now to this problem.

The interaction potential is assumed to be expressible in a form which is analogous to equation (5.41), namely

$$V(\hat{\mathbf{r}}'_1, \hat{\mathbf{r}}'_2, R) = \sum_{\lambda_1 \lambda_2 \mu \geq 0} v_{\lambda_1 \lambda_2 \mu}(R) Y_{\lambda_1 \lambda_2 \mu}(\hat{\mathbf{r}}'_1, \hat{\mathbf{r}}'_2) \tag{6.17}$$

where

$$Y_{\lambda_1 \lambda_2 \mu}(\hat{\mathbf{r}}'_1, \hat{\mathbf{r}}'_2) = \frac{4\pi}{[2(1 + \delta_{\mu 0})]^{\frac{1}{2}}}$$

$$\times [Y_{\lambda_1 \mu}(\hat{\mathbf{r}}'_1) Y_{\lambda_2, -\mu}(\hat{\mathbf{r}}'_2) + Y_{\lambda_1, -\mu}(\hat{\mathbf{r}}'_1) Y_{\lambda_2 \mu}(\hat{\mathbf{r}}'_2)] \tag{6.18}$$

and where $\hat{\mathbf{r}}'_1 = (\theta'_1, \phi'_1)$ and $\hat{\mathbf{r}}'_2 = (\theta'_2, \phi'_2)$ denote the internal angular coordinates of systems 1 and 2 in the BF frame. The interaction potentials between linear molecules in Σ electronic

6.2 Theory of fine structure excitation processes

states (e.g. H$_2$–H$_2$) or between such a molecule and an atom with an open shell may be expressed in this way. The matrix elements of V may then be written as

$$\langle \gamma j_{12} \bar{\Omega} p J M | V | \gamma' j'_{12} \bar{\Omega}' p' J' M' \rangle = \delta_{pp'} \delta_{JJ'} \delta_{MM'}$$

$$\times \sum_{\lambda_1 \lambda_2 \mu \geq 0} v_{\lambda_1 \lambda_2 \mu}(R) \langle \gamma j_{12} \bar{\Omega} p J M | Y_{\lambda_1 \lambda_2 \mu}(\hat{\mathbf{r}}'_1, \hat{\mathbf{r}}'_2) | \gamma' j'_{12} \bar{\Omega}' p' J' M' \rangle \quad (6.19)$$

and evaluated using standard techniques of angular momentum ('Racah') algebra. Assuming that the angular momenta, $\mathbf{j}_i$ ($i = 1, 2$), of the two systems, 1 and 2, are composed of orbital and spin angular momenta, that is

$$\mathbf{j}_i = \mathbf{l}_i + \mathbf{s}_i$$

and defining

$$\lambda_{12} = \lambda_1 + \lambda_2$$

it may be shown [119] that

$$\langle \gamma j_{12} \bar{\Omega} p J M | Y_{\lambda_1 \lambda_2 \mu}(\hat{\mathbf{r}}'_1, \hat{\mathbf{r}}'_2) | \gamma' j'_{12} \bar{\Omega}' p J M \rangle$$

$$= \delta_{\bar{\Omega}\bar{\Omega}'} \delta_{s_1 s'_1} \delta_{s_2 s'_2} (-1)^{s_1 + j'_1 + s_2 + j'_2 + j_{12} - \bar{\Omega}} \left(\frac{2}{1 + \delta_{\mu 0}} \right)^{\frac{1}{2}}$$

$$\times [(2j_1 + 1)(2j_2 + 1)(2j_{12} + 1)(2j'_1 + 1)(2j'_2 + 1)(2j'_{12} + 1)$$

$$\times (2l_1 + 1)(2\lambda_1 + 1)(2l'_1 + 1)(2l_2 + 1)(2\lambda_2 + 1)(2l'_2 + 1)]^{\frac{1}{2}}$$

$$\times \begin{pmatrix} l_1 & \lambda_1 & l'_1 \\ 0 & 0 & 0 \end{pmatrix} \begin{pmatrix} l_2 & \lambda_2 & l'_2 \\ 0 & 0 & 0 \end{pmatrix}$$

$$\times \begin{Bmatrix} l_1 & j_1 & s_1 \\ j'_1 & l'_1 & \lambda_1 \end{Bmatrix} \begin{Bmatrix} l_2 & j_2 & s_2 \\ j'_2 & l'_2 & \lambda_2 \end{Bmatrix} \sum_{\lambda_{12}} (2\lambda_{12} + 1) \begin{pmatrix} \lambda_1 & \lambda_2 & \lambda_{12} \\ -\mu & \mu & 0 \end{pmatrix}$$

$$\times \begin{pmatrix} j_{12} & \lambda_{12} & j'_{12} \\ \bar{\Omega} & 0 & \bar{\Omega} \end{pmatrix} \begin{Bmatrix} j_1 & j_2 & j_{12} \\ j'_1 & j'_2 & j'_{12} \\ \lambda_1 & \lambda_2 & \lambda_{12} \end{Bmatrix} \quad (6.20)$$

In (6.20) Wigner 3j-, 6j-, and 9j-coefficients appear [51]. Once again, the matrix elements of both $\mathbf{l}^2$ and V are independent of the projection, M, of the total angular momentum, $\mathbf{J}$, on the SF z-axis; the orientation of the z-axis may be chosen arbitrarily. It follows that $M (= M')$ may be dropped from the notation, and the coupled equations (6.14) take the form

$$\left[\frac{d^2}{dR^2} + k_\gamma^2 \right] G(\gamma j_{12} \bar{\Omega} p J | R)$$

$$= 2\mu \sum_{\gamma' j'_{12} \bar{\Omega}'} V_{\text{eff}}(\gamma j_{12} \bar{\Omega}, \gamma' j'_{12} \bar{\Omega}'; p J) G(\gamma' j'_{12} \bar{\Omega}' p J | R) \quad (6.21)$$

where $V_{\text{eff}}(\gamma j_{12}\bar{\Omega}, \gamma' j'_{12}\bar{\Omega}'; pJ)$ is a matrix element of the effective potential $V_{\text{eff}} = V + \mathbf{l}^2/(2\mu R^2)$.

The equations (6.21) may be solved using techniques analogous to those which were considered in Chapter 4, applicable when the following boundary conditions apply:

$$V \to \infty \quad (R \to 0)$$

and

$$V \sim R^{-n} \quad (R \to \infty)$$

where the integer $n \geq 2$. A problem involving N coupled equations has a total of N linearly independent solution vectors (corresponding to the different possible initial conditions), each comprising N elements (corresponding to the different possible final conditions). If the solution vectors are written as the columns of an $N \times N$ matrix, $\mathbf{G}(R)$, then the coupled equations (6.21) may be expressed in the compact matrix form

$$\left[\mathbf{1}\frac{d^2}{dR^2} + \mathbf{W}(R)\right]\mathbf{G}(R) = 0 \tag{6.22}$$

where $\mathbf{1}$ denotes the unit matrix and

$$W(\gamma j_{12}\bar{\Omega}, \gamma' j'_{12}\bar{\Omega}'; pJ|R)$$
$$= \delta_{\gamma\gamma'}\delta_{j_{12}j'_{12}}\delta_{\bar{\Omega}\bar{\Omega}'}k_\gamma^2 - 2\mu V_{\text{eff}}(\gamma j_{12}\bar{\Omega}, \gamma' j'_{12}\bar{\Omega}'; pJ|R) \tag{6.23}$$

The solutions of the BF-coupled equations may be propagated numerically into the asymptotic region, starting from the classically forbidden region, close to the origin, where

$$\mathbf{G}(R) = \mathbf{0} \tag{6.24}$$

and $\mathbf{0}$ is the null matrix. In the asymptotic region, where V has become vanishingly small (in practice, small compared with the collision energy), a transformation to SF coordinates may be performed, using equations analogous to (4.28) and (4.29), namely

$$|\gamma j_{12}lpJ\rangle = |\gamma j_{12}\bar{\Omega}pJ\rangle\langle\gamma j_{12}\bar{\Omega}pJ|\gamma j_{12}lpJ\rangle \tag{6.25}$$

where there is an implied summation over $\bar{\Omega}$ and

$$\langle\gamma j_{12}\bar{\Omega}pJ|\gamma j_{12}lpJ\rangle = \left[\frac{2(2l+1)}{(1+\delta_{\bar{\Omega}0})(2J+1)}\right]^{\frac{1}{2}} C^{j_{12}lJ}_{\bar{\Omega}0\bar{\Omega}} \tag{6.26}$$

which is obtained by setting $p = p'$ in equation (6.9). The information on the scattering event can then be extracted by matching the resulting SF wave functions, $F(\gamma j_{12}lpJ|R)$, to the appropriate asymptotic $(R \to \infty)$ forms

$$\mathbf{F}(R) \sim \mathbf{J}(R)\mathbf{A} - \mathbf{N}(R)\mathbf{B} \tag{6.27}$$

6.2 Theory of fine structure excitation processes

where $\mathbf{J}(R)$ and $\mathbf{N}(R)$ are diagonal matrices with elements

$$J(\gamma j_{12}l, \gamma' j'_{12}l'; pJ|R) = \delta_{\gamma\gamma'}\delta_{j_{12}j'_{12}}\delta_{ll'}k_\gamma^{\frac{1}{2}} R\, j_l(k_\gamma R) \tag{6.28}$$

and

$$N(\gamma j_{12}l, \gamma' j'_{12}l'; pJ|R) = \delta_{\gamma\gamma'}\delta_{j_{12}j'_{12}}\delta_{ll'}k_\gamma^{\frac{1}{2}} R\, n_l(k_\gamma R) \tag{6.29}$$

j_l and n_l are the spherical Bessel functions of the first and second kinds, respectively [47]. In order to accommodate *closed channels* in the basis of scattering functions, i.e. states of the system that are not energetically accessible at infinite separation of the projectile and target, spherical Bessel functions of the third kind are also required. These functions decay exponentially with R as $R \to \infty$. The inclusion of some closed channels can be essential to obtaining accurate numerical results, particularly for transitions involving states whose energies approach the limit of accessibility at infinite separation.

The reactance matrix $\mathbf{K}$ is given by

$$\mathbf{K} = \mathbf{B}\mathbf{A}^{-1} \tag{6.30}$$

and is related to the *transmission* matrix $\mathbf{T}$ and the scattering matrix $\mathbf{S}$ through

$$\mathbf{T} = -2i\mathbf{K}(1 - i\mathbf{K})^{-1}$$
$$\equiv 1 - \mathbf{S} \tag{6.31}$$

The probability P of the transition $\gamma' \to \gamma$ is

$$P_{Jp}(\gamma \leftarrow \gamma') = \frac{1}{(2j'_1+1)(2j'_2+1)} \sum_{\substack{j_{12}l \\ j'_{12}l'}} |T(\gamma j_{12}l, \gamma' j'_{12}l'; pJ)|^2 \tag{6.32}$$

for each value of the total angular momentum J and of the parity p. The total cross-section is

$$\sigma(\gamma \leftarrow \gamma') = \frac{\pi}{k_{\gamma'}^2} \sum_{Jp}(2J+1) P_{Jp}(\gamma \leftarrow \gamma') \tag{6.33}$$

Excitation of C^+ $2p\,^2P^o$ in collisions with H_2

As a first illustration of the use of the above formalism, we consider the process

$$C^+(2p^2P^o_{\frac{1}{2}}) + H_2 \to C^+(2p^2P^o_{\frac{3}{2}}) + H_2$$

Let us denote the H_2 molecule, which is assumed to remain in its ground electronic and vibrational states, as system 1, and the C^+ ion as system 2. Then, $s_1 = 0$ and $j_1 = l_1$ is the rotational angular momentum of the H_2 molecule. Furthermore, $s_2 = s'_2 = \frac{1}{2}$ and $l_2 = l'_2 = 1$, and hence $j_2, j'_2 = \frac{1}{2}, \frac{3}{2}$, corresponding to the two fine structure levels of the ^2P ground term of C^+.

The interaction potential (6.17) may be derived by considering the symmetry constraints on the wave function of the 2p electron of the C^+ ion as it approaches the H_2 molecule. Group theory provides the framework and the notation for the associated potential energy curves. A brief but characteristically excellent summary of the classification of the electronic states of non-linear molecules has been given by Herzberg [120].

For approaches of the ion *perpendicular to the internuclear axis of the molecule*, the appropriate point group is C_{2v} and the three lowest adiabatic potential energy curves, which correlate at infinite separation with H_2 and C^+ in their electronic ground states, are denoted $^2A_1$, $^2B_1$ and $^2B_2$. The electronic orbitals of appropriate symmetry are

$$|^2A_1\rangle = Y_{10}(\hat{\mathbf{r}}'_2) \propto \cos(\theta'_2)$$

$$|^2B_1\rangle = \frac{[Y_{11}(\hat{\mathbf{r}}'_2) - Y_{1,-1}(\hat{\mathbf{r}}'_2)]}{2^{\frac{1}{2}}} \propto \sin(\theta'_2)\cos(\phi'_2)$$

$$|^2B_2\rangle = \frac{[Y_{11}(\hat{\mathbf{r}}'_2) + Y_{1,-1}(\hat{\mathbf{r}}'_2)]}{2^{\frac{1}{2}}} \propto \sin(\theta'_2)\sin(\phi'_2) \qquad (6.34)$$

The functions on the right-hand sides of (6.34) are linear combinations of the spherical harmonics representing the three possible angular momentum states, $Y_{lm_l}(\hat{\mathbf{r}}'_2)$, of the valence p (i.e. $l=1$) electron. The functions (6.34) are eigenfunctions of the operators of the group: the identity operator; the operator that effects a reflection in the symmetry plane of the system, which is defined by the internuclear axis and the line joining the ion to the centre of mass of the molecule (which is located at the mid-point of the intramolecular axis); and the operator that effects a rotation through $2\pi/2 = \pi$ about the the line joining the ion to the centre of mass of the molecule. The three orbitals in equation (6.34) may be denoted p_Z, p_X and p_Y, respectively, as their lobes are orientated along the Z-, X- and Y-axes, as may be seen from their dependence on θ'_2 and ϕ'_2. The adiabatic potential energy curves are determined by

$$V(^2A_1) = \langle ^2A_1|V(\theta'_1 = \pi/2 = \phi'_1, \hat{\mathbf{r}}'_2, R)|^2A_1\rangle_{\hat{\mathbf{r}}'_2}$$

$$V(^2B_1) = \langle ^2B_1|V(\theta'_1 = \pi/2 = \phi'_1, \hat{\mathbf{r}}'_2, R)|^2B_1\rangle_{\hat{\mathbf{r}}'_2}$$

$$V(^2B_2) = \langle ^2B_2|V(\theta'_1 = \pi/2 = \phi'_1, \hat{\mathbf{r}}'_2, R)|^2B_2\rangle_{\hat{\mathbf{r}}'_2} \qquad (6.35)$$

In (6.35), the integration is over the electronic coordinates $\hat{\mathbf{r}}'_2 = (\theta'_2, \phi'_2)$, for fixed values of the H_2 internuclear coordinates $\hat{\mathbf{r}}'_1 = (\pi/2, \pi/2)$ (all in the BF system: see Fig. 6.1). The latter are determined by the requirement that, in C_{2v} symmetry, the approach of the C^+ ion should be perpendicular to the internuclear axis of the H_2 molecule, and hence $\theta'_1 = \pi/2$, and by the convention which we adopt that the symmetry plane, containing all three nuclei, should be the YZ-plane, and hence $\phi'_1 = \pi/2$. The use of the Born–Oppenheimer approximation is implicit in equations (6.35), which are expressions for the 'adiabatic' potentials, obtained as expectation values with respect to the electronic coordinates for *fixed* values of the H_2 internuclear coordinates.

For *collinear* approaches of the C^+–H–H system, the relevant point group is $C_{\infty v}$. In this case, the adiabatic potential energy curves are

$$V(^2\Sigma) = \langle ^2\Sigma|V(\theta'_1 = 0, \phi'_1 = \pi/2, \hat{\mathbf{r}}'_2, R)|^2\Sigma\rangle_{\hat{\mathbf{r}}'_2}$$

$$V(^2\Pi) = \langle ^2\Pi^\pm|V(\theta'_1 = 0, \phi'_1 = \pi/2, \hat{\mathbf{r}}'_2, R)|^2\Pi^\pm\rangle_{\hat{\mathbf{r}}'_2} \qquad (6.36)$$

6.2 Theory of fine structure excitation processes

Table 6.2. *Form of the long-range interaction between an atomic ion such as C^+ and a homonuclear diatomic molecule such as H_2. The quadrupole moment of the molecule is denoted Θ, and $\alpha = (\alpha_\| + 2\alpha_\perp)/3$ is its mean polarizability, where $\alpha_\|$ and $\alpha_\perp$ are the polarizabilities parallel and perpendicular to its internuclear axis, respectively; $\Delta\alpha = \alpha_\| - \alpha_\perp$; $\langle r_2^2 \rangle$ is the mean square radius of the 2p valence electron of the C^+ ion. Numerical values (in atomic units) of these constants are: $\Theta = 0.46$ [123], $\alpha_\| = 6.38$, $\alpha_\perp = 4.58$, $\alpha = 5.18$ [124], and $\langle r_2^2 \rangle = 2.689$ [125].*

λ_1	λ_2	μ			
0	0	0		$-\alpha/(2R^4)$	$-\alpha\langle r_2^2\rangle/R^6$
2	0	0	$\Theta/(5^{\frac{1}{2}}R^3)$	$-\Delta\alpha/[3((5^{\frac{1}{2}})R^4]$	$-\Delta\alpha\langle r_2^2\rangle/[3(5^{\frac{1}{2}})R^6]$
0	2	0			$2\alpha\langle r_2^2\rangle/(5^{\frac{1}{2}}R^6)$
2	2	0		$-6\Theta\langle r_2^2\rangle/(5R^5)$	$\Delta\alpha\langle r_2^2\rangle/(5R^6)$
2	2	1		$-4(2^{\frac{1}{2}})\Theta\langle r_2^2\rangle/(5R^5)$	$2^{\frac{1}{2}}\Delta\alpha\langle r_2^2\rangle/(15R^6)$
2	2	2		$-2^{\frac{1}{2}}\Theta\langle r_2^2\rangle/(5R^5)$	$-2^{\frac{1}{2}}\Delta\alpha\langle r_2^2\rangle/(15R^6)$

where

$$|^2\Sigma\rangle = Y_{10}(\hat{\mathbf{r}}_2')$$

$$|^2\Pi^\pm\rangle = \frac{[Y_{11}(\hat{\mathbf{r}}_2') \pm Y_{1,-1}(\hat{\mathbf{r}}_2')]}{2^{\frac{1}{2}}} \quad (6.37)$$

The $^2\Pi^+$ and $^2\Pi^-$ states are degenerate (have the same energy) and may be denoted simply by $^2\Pi$.

Carrying out the integration with respect to $\hat{\mathbf{r}}_2'$ in equations (6.35) and (6.36) yields expressions for the five adiabatic potential energy curves ($^2A_1$, $^2B_1$, $^2B_2$, $^2\Sigma$, $^2\Pi$) in terms of the five coefficients ($v_{000}, v_{200}, v_{020}, v_{220}, v_{222}$) of the potential energy expansion (6.17) [121]. Inversion of these linear algebraic equations enables the coefficients $v_{\lambda_1\lambda_2\mu}(R)$ to be determined from the potential energy curves, computed as functions of R. In order to determine the coefficient $v_{221}(R)$, the potential energy curves must also be known for geometries intermediate between the perpendicular and collinear approaches (i.e. for the general C_s symmetry). At long-range, explicit relations for all six coefficients may be derived from perturbation theory [122]; these are given in Table 6.2. In principle, the results obtained from the potential energy curves calculated at short and intermediate range should join the long-range forms smoothly. In practice, a smooth transition may not be realized and has to be imposed. If this is not done, the quantum mechanical scattering calculations predict (non-physical) reflections at the associated steps in the interaction potential, which affect the values of the cross-sections.

In order to calculate reliably the cross-section over a wide range of collision energies, the interaction potential must be known at short and intermediate as well as long range; this necessitates accurate *ab initio* computations of the adiabatic potential energy curves. However, in the context of the cooling of the interstellar medium by fine structure transitions, the behaviour of the cross-section at low collision energies is often the most important consideration. As the collision energy falls, the significant part of the interaction potential tends to move to larger values of R. The long-range form of the potential (Table 6.2) can then provide a reliable guide to what happens during the scattering process.

Interesting and significant orientation-dependent effects occur in C^+–H_2 collisions at low energies. The C^+ ion prefers to approach the H_2 molecule *perpendicular* to its internuclear axis, as the long-range interaction potential for this geometry is *attractive*:

$$V_\perp(R) = -\frac{\Theta}{2R^3} - \frac{\alpha_\perp}{2R^4} \tag{6.38}$$

In the collinear approach, on the other hand, the potential

$$V_\parallel(R) = \frac{\Theta}{R^3} - \frac{\alpha_\parallel}{2R^4} \tag{6.39}$$

presents a barrier whose height is about 50 K (at $R \approx 9\ a_0$). [In equations (6.38, 6.39), Θ is the quadrupole moment of the H_2 molecule and $\alpha_\parallel$, $\alpha_\perp$ are the polarizabilities parallel and perpendicular to its internuclear axis, respectively.] At low energies, this barrier prevents the ion approaching sufficiently closely to the molecule for the fine structure transition to be induced. For this reason, it is essential to consider the potential energy curves in C_{2v} symmetry as well as in $C_{\infty v}$ symmetry, as discussed above.

A simple interpretation of the fine structure excitation process is suggested by plots of the excitation probability, $P_J(\frac{3}{2} \leftarrow \frac{1}{2})$, against the total angular momentum, J. Such a plot is shown in Fig. 6.2 for collisions of C^+ with para-H_2, at a collision energy $E/k_B = 180$ K.

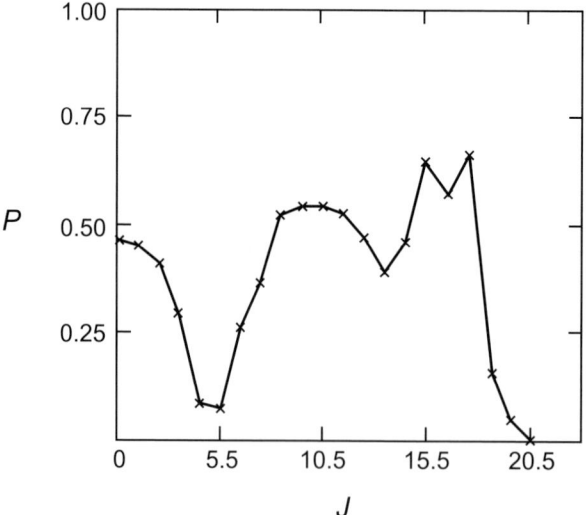

Figure 6.2 The probability, P_J, of the $^2P_{\frac{1}{2}} \to {}^2P_{\frac{3}{2}}$ transition in C^+, induced by collisions with para-H_2; J is the total angular momentum of the C^+–H_2 system. The centre-of-mass collision energy is $E/k_B = 180$ K.

6.2 Theory of fine structure excitation processes

The figure shows that $P_J \approx 0.5$ for $J \leq 17.5$, whereas $P_J \approx 0.0$ for $J > 17.5$. As we shall now see, $J = 17.5$ corresponds closely to the classical impact parameter for *orbiting*.

Consider the C^+ ion moving relative to the H_2 molecule in an effective potential determined by the polarization attraction and the centrifugal repulsion

$$V_{\text{eff}}(R) = -\frac{\alpha}{2R^4} + \frac{Eb^2}{R^2} \tag{6.40}$$

where α is the mean polarizability of the H_2 molecule, E is the collision energy and b denotes the impact parameter. The effective potential is repulsive at long range, and attractive at short range; it has a maximum for $R = (\alpha/E)^{\frac{1}{2}}/b$ of magnitude $E^2 b^4/(2\alpha)$. Excitation occurs when the collision energy, E, is just sufficient to overcome the effective potential barrier and the excitation energy, Δ, of the transition, that is, when

$$E = \frac{E^2 b^4}{2\alpha} + \Delta \tag{6.41}$$

or

$$b^4 = \frac{2\alpha(E - \Delta)}{E^2} \tag{6.42}$$

The orbiting radius, $R = (\alpha/E)^{\frac{1}{2}}/b$, is a point of unstable equilibrium at which the radial component of the relative velocity of the ion–molecule pair is zero. To the critical impact parameter, determined by equation (6.42), there corresponds a value of the relative angular momentum

$$J^2 = 2\mu E b^2 = 2\mu [2\alpha(E - \Delta)]^{\frac{1}{2}} \tag{6.43}$$

where μ is the reduced mass of the ion–molecule system. In the case of C^+–H_2 and for $E/k_B = 180$ K, equation (6.43) yields $J \approx 18.5$, which agrees well with the value at the cut-off in Fig. 6.2.

If $J \lesssim 18.5$, the centrifugal repulsion is insufficient to prevent the ions penetrating close enough to the molecule for the fine structure transition to occur, with a probability $P_J \approx 0.5$, as seen in Fig. 6.2. The corresponding expression for the cross-section is

$$\sigma\left(\frac{3}{2} \leftarrow \frac{1}{2}\right) = \frac{\pi [2\alpha(E - \Delta)]^{\frac{1}{2}}}{2E} \tag{6.44}$$

where $\alpha = 5.18$ is the mean polarizability of H_2 and $\Delta = 91$ K. In Table 6.3 are compared the predictions of equation (6.44) with the results of quantum mechanical calculations of $\sigma(3/2 \leftarrow 1/2)$. The good agreement attests to the basic validity of the semi-classical model from which equation (6.44) derives.

The results in Table 6.3 for para-H_2 were obtained with the basis of rotational states $j_1 = 0, 2$, and those for ortho-H_2 with the basis $j_1 = 1, 3$. The cross-section is larger in the case of ortho- than in the case of para-H_2 because of the greater importance of interactions

Table 6.3. *Cross-sections, $\sigma(\frac{3}{2} \leftarrow \frac{1}{2})$ (in atomic units, a_0^2), for excitation of the C^+ $2p\ ^2P^o_{\frac{1}{2}} \rightarrow 2p^2P^o_{\frac{3}{2}}$ fine structure transition by para-H_2 and ortho-H_2 [126], as a function of the centre-of-mass collision energy, E. The last column contains, for comparison, results obtained using the semi-classical expression [equation (6.44)].*

E (K)	para-H_2	ortho-H_2	Equation (6.44)
94	34.5	55.4	42.7
97	46.8	82.2	65.5
100	67.3	98.3	80.4
105	84.2	114	97.6
112	98.8	135	113
120	103	149	125
135	147	173	138
150	145	187	144
180	152	191	148
210	158	188	147
250	168	194	143
300	172	186	137
360	166	180	129
440	145	184	120
550	131	172	111
750	122	165	97.2

involving the quadrupole moment of the molecule. These interactions are neglected by the semi-classical model, which in effect treats the molecule as a spherically symmetric system.

Excitation of C $2p^2$ ^{3}P and O $2p^4$ ^{3}P in collisions with H_2

C $2p^2$ ^{3}P has two valence electrons, whereas O $2p^4$ ^{3}P has two vacancies ('holes') in the valence shell; both atoms have total orbital angular momentum $L=1$ and total spin $S=1$ in their ground ^{3}P terms. The (triplet) spin state is symmetric under electron exchange, and hence the spatial part of the wave function must be asymmetric, to satisfy the Pauli exclusion principle (Fermi–Dirac statistics). The angular part of the electronic wave function, which determines its spatial symmetry properties, is given by

$$Y_{1M}(\hat{\mathbf{r}}'_2) = \sum_{m_1 m_2} C^{111}_{m_1 m_2 M} Y_{1m_1}(\hat{\rho}'_1) Y_{1m_2}(\hat{\rho}'_2) \tag{6.45}$$

In this equation, $\hat{\rho}'_1$ and $\hat{\rho}'_2$ denote the angular coordinates of the two valence electrons (or valence electron holes), and $\hat{\mathbf{r}}'_2$ represents the angular coordinates of both electrons, all in the BF coordinate system. The total angular momentum projection quantum number $M = m_1 + m_2 = -1, 0$ or 1. Substituting the values of the Clebsch–Gordan coefficients,

6.2 Theory of fine structure excitation processes

$C^{111}_{m_1 m_2 M}$, one obtains

$$Y_{11}(\hat{\mathbf{r}}'_2) = -2^{-\frac{1}{2}}[Y_{10}(\hat{\rho}'_1)Y_{11}(\hat{\rho}'_2) - Y_{11}(\hat{\rho}'_1)Y_{10}(\hat{\rho}'_2)]$$

$$Y_{10}(\hat{\mathbf{r}}'_2) = 2^{-\frac{1}{2}}[Y_{11}(\hat{\rho}'_1)Y_{1,-1}(\hat{\rho}'_2) - Y_{1,-1}(\hat{\rho}'_1)Y_{11}(\hat{\rho}'_2)]$$

$$Y_{1,-1}(\hat{\mathbf{r}}'_2) = 2^{-\frac{1}{2}}[Y_{10}(\hat{\rho}'_1)Y_{1,-1}(\hat{\rho}'_2) - Y_{1,-1}(\hat{\rho}'_1)Y_{10}(\hat{\rho}'_2)] \quad (6.46)$$

It may be seen that the functions (6.46) are asymmetric under exchange of the electron coordinates, $\hat{\rho}'_1$ and $\hat{\rho}'_2$, as required.

Consider now C_{2v} symmetry, in which the C (or O) atom approaches the H$_2$ molecule perpendicular to its internuclear axis. In this case, the three lowest adiabatic potential energy curves are denoted $^3A_2$, $^3B_1$ and $^3B_2$, and the normalized electronic eigenfunctions of appropriate symmetry are

$$|^3A_2\rangle = 2^{-\frac{1}{2}}[Y_{11}(\hat{\rho}'_1)Y_{1,-1}(\hat{\rho}'_2) - Y_{1,-1}(\hat{\rho}'_1)Y_{11}(\hat{\rho}'_2)]$$

$$|^3B_1\rangle = 2^{-1}[Y_{10}(\hat{\rho}'_1)Y_{11}(\hat{\rho}'_2) - Y_{11}(\hat{\rho}'_1)Y_{10}(\hat{\rho}'_2)]$$
$$- 2^{-1}[Y_{10}(\hat{\rho}'_1)Y_{1,-1}(\hat{\rho}'_2) - Y_{1,-1}(\hat{\rho}'_1)Y_{10}(\hat{\rho}'_2)]$$

$$|^3B_2\rangle = 2^{-1}[Y_{10}(\hat{\rho}'_1)Y_{11}(\hat{\rho}'_2) - Y_{11}(\hat{\rho}'_1)Y_{10}(\hat{\rho}'_2)]$$
$$+ 2^{-1}[Y_{10}(\hat{\rho}'_1)Y_{1,-1}(\hat{\rho}'_2) - Y_{1,-1}(\hat{\rho}'_1)Y_{10}(\hat{\rho}'_2)] \quad (6.47)$$

Comparing equations (6.46) and (6.47), we see that

$$|^3A_2\rangle = Y_{10}(\hat{\mathbf{r}}'_2)$$

$$|^3B_1\rangle = -\frac{[Y_{11}(\hat{\mathbf{r}}'_2) + Y_{1,-1}(\hat{\mathbf{r}}'_2)]}{2^{\frac{1}{2}}}$$

$$|^3B_2\rangle = -\frac{[Y_{11}(\hat{\mathbf{r}}'_2) - Y_{1,-1}(\hat{\mathbf{r}}'_2)]}{2^{\frac{1}{2}}} \quad (6.48)$$

If these expressions are compared with equation (6.34), it may be seen that there is an equivalence between the adiabatic potential energy curves $V(^3A_2)$, $V(^3B_1)$ and $V(^3B_2)$ in the two-electron case with $V(^2A_1)$, $V(^2B_2)$ and $V(^2B_1)$, respectively, in the one-electron case. [Note that the overall sign of the electronic eigenfunction is not significant in the present context.] The corresponding analysis for $C_{\infty v}$ symmetry, i.e. for the collinear approach of the atom to the molecule, demonstrates the equivalence of $V(^3\Sigma^-)$ and $V(^3\Pi)$ to $V(^2\Sigma)$ and $V(^2\Pi)$ [equation (6.36)]. Subject to these equivalences, the expansion coefficients $v_{\lambda_1 \lambda_2 \mu}(R)$

of the interaction potential (6.17) may be derived from the C–H$_2$ (and O–H$_2$) adiabatic potential energy curves, computed for the collinear and perpendicular approaches.

6.2.2 Systems involving more than one open shell

The formulation above is not readily extended to collisions between two systems, both with open shells, where at least two active electrons are involved and electron-exchange effects become important. Launay and Roueff [127, 128] presented a formulation of the problem of fine structure excitation in collisions between open-shell systems that owes much to earlier work by Wofsy et al. [129]; this theory will now be outlined.

The Schrödinger equation describing the interacting particles may be written, in the SF frame, in the form

$$\left(\frac{d^2}{dR^2} + k_\gamma^2\right) F(\gamma j_{12} l p J | R)$$

$$= 2\mu \sum_{\gamma' j'_{12} l'} V_{\text{eff}}(\gamma j_{12} l, \gamma' j'_{12} l'; p J | R) F(\gamma' j'_{12} l' p J | R) \tag{6.49}$$

where $V_{\text{eff}}(\gamma j_{12} l, \gamma' j'_{12} l'; p J | R)$ is a matrix element of the effective potential, $V_{\text{eff}} = V + \mathbf{l}^2/(2\mu R^2)$. In the SF representation, the matrix elements of the angular momentum operator, $\mathbf{l}^2$, take the simple form

$$\langle \gamma j_{12} l p J | \mathbf{l}^2 | \gamma' j'_{12} l' p J \rangle = \delta_{\gamma \gamma'} \delta_{j_{12} j'_{12}} \delta_{l l'} l(l+1) \tag{6.50}$$

in which case equation (6.49) may be written as

$$\left[\frac{d^2}{dR^2} - \frac{l(l+1)}{R^2} + k_\gamma^2\right] F(\gamma j l p J | R)$$

$$= 2\mu \sum_{\gamma' j'_{12} l'} \langle \gamma j_{12} l p J | V | \gamma' j'_{12} l' p J \rangle F(\gamma' j'_{12} l' p J | R) \tag{6.51}$$

It may be seen from the analysis of collisions between systems with one open shell, presented above, that the matrix elements of the potential, $V(R)$, can be written in terms of linear combinations of the relevant adiabatic potential energy curves. If we consider interactions between atoms or between an atom and an ion, the collision system is diatomic and the adiabatic potentials may be denoted $V_{\Lambda,S}$, where $\Lambda = |M_L|$ and $M_L = M_{L_1} + M_{L_2}$ is the sum of the projections of the electronic *orbital* angular momenta on the *internuclear* axis; S is the resultant spin angular momentum quantum number, i.e. $\mathbf{S} = \mathbf{S}_1 + \mathbf{S}_2$. When one of the collision partners (2, say) is in a state of zero orbital angular momentum ($L_2 = 0 = M_{L_2}$), as is the case of H(1s ^{2}S), the explicit expressions for the matrix elements of the potential are [128]

6.2 Theory of fine structure excitation processes

$$\langle \gamma j_{12} l p J | V | \gamma' j'_{12} l' p J \rangle = \sum_{\Lambda,S} (-1)^{J+S+j_{12}+j'_{12}} V_{\Lambda,S}(R)$$

$$\times \sum_{|M_L|=\Lambda} (2S+1)[(2j_1+1)(2j'_1+1)(2j_{12}+1)(2j'_{12}+1)(2l+1)(2l'+1)]^{\frac{1}{2}}$$

$$\times \begin{Bmatrix} L_1 & j_{12} & S \\ S_2 & S_1 & j_1 \end{Bmatrix} \begin{Bmatrix} L_1 & j'_{12} & S \\ S_2 & S_1 & j'_1 \end{Bmatrix} \sum_c \begin{Bmatrix} L_1 & L_1 & c \\ j_{12} & j'_{12} & S \end{Bmatrix} \begin{Bmatrix} l' & l & c \\ j_{12} & j'_{12} & J \end{Bmatrix}$$

$$\times \begin{pmatrix} l' & l & c \\ 0 & 0 & 0 \end{pmatrix} \begin{pmatrix} L_1 & L_1 & c \\ -M_L & M_L & 0 \end{pmatrix} (-1)^{M_L}(2c+1) \qquad (6.52)$$

where c is the order of the potential coupling.

Solving the coupled equations (6.51) yields the scattering and transmission matrices. The transition probability is given in terms of the elements of the transmission matrix by

$$P_{Jp}(\gamma \leftarrow \gamma') = \frac{1}{(2j'_1+1)(2S_2+1)} \sum_{\substack{j_{12}l \\ j'_{12}l'}} |T(\gamma j_{12}l, \gamma' j'_{12}l'; pJ)|^2 \qquad (6.53)$$

and the total cross-section by

$$\sigma(\gamma \leftarrow \gamma') = \frac{\pi}{k_{\gamma'}^2} \sum_{Jp} (2J+1) P_{Jp}(\gamma \leftarrow \gamma') \qquad (6.54)$$

Excitation of C^+ 2p $^2P^o$ in collisions with H

In this case, $L_1 = 1, L_2 = 0, S_1 = \frac{1}{2} = S_2$, and $j_1 = \frac{1}{2}, \frac{3}{2}$, where the subscript 1 denotes the C^I ion and 2 the H atom. The possible values of both Λ and S are 0 and 1, to which correspond the four adiabatic potential energy curves $^1\Sigma$, $^3\Sigma$, $^1\Pi$ and $^3\Pi$, which correlate with C^+ 2p $^2P^o + H^0(1s\ ^2S)$ as $R \to \infty$. Launay and Roueff [127] used the adiabatic potential energy curves calculated out to $R = 6.5 a_0$ by Green et al. [130], together with their known long-range forms, to compute the fine structure excitation cross-sections in Table 6.4.

Wofsy et al. [129] used a simpler model of the collision process, involving the evaluation of elastic scattering phase shifts. As may be seen from Table 6.4, the approximation of neglecting the inelasticity, that is, the energy difference between the fine structure levels (also known as the energy defect), is least accurate near threshold (91 K) and improves with increasing collision energy, E, as might have been anticipated.

The semi-classical orbiting approximation, considered above in connection with C^+–H_2 scattering, may be applied here also. The calculations of Launay and Roueff [127] suggest that the excitation probability, $P_J \approx 2/3$, corresponding to the ratio of the statistical weight of the upper fine structure level (4) and the statistical weight of the doublet (6). The corresponding expression for the cross-section is

$$\sigma(\frac{3}{2} \leftarrow \frac{1}{2}) = \frac{2\pi[2\alpha(E-\Delta)]^{\frac{1}{2}}}{3E} \qquad (6.55)$$

Table 6.4. *Cross-sections, $\sigma(\frac{3}{2} \leftarrow \frac{1}{2})$ (in atomic units, a_0^2), for excitation of the C^+ 2p $^2P^o_{\frac{1}{2}} \to 2p\ ^2P^o_{\frac{3}{2}}$ fine structure transition by $H^0(1s\ ^2S)$, as a function of the centre of mass collision energy, E. (a) and (b) are taken from [127], where the fine structure energy defect ($\Delta = 92$ K in their calculations) is neglected in (b); (c) is from the orbiting approximation (see text).*

E (K)	a	b	c
93	41.6		38.0
95	70.3		64.4
100	109	382	100
125	178		162
150	220		179
175	232		184
200	229	285	183
250	219		178
300	213	239	170
400	200	201	155
600	170	174	133
800	158	157	120
1000	145	149	106
3000	109		63.5

where $\alpha = 4.5$ is the polarizability of H^0 and Δ is the fine structure energy defect. Results obtained using equation (6.55) are listed in the last column of Table 6.4. Once again, the level of agreement testifies to the validity of the orbiting model.

The sensitivity of the cross-section to uncertainties in the interaction potential was considered by both Launay and Roueff [127] and by Harel *et al.* [131]. Launay and Roueff used two forms of the potential, both of which were consistent with the computed potential, to within the uncertainties. The results of the two calculations of the cross-section differed by up to 40 per cent, which may still be considered acceptable for many astrophysical applications. The conclusions of Harel *et al.* were consistent with those that may be drawn from the calculations of Launay and Roueff. This problem needs to be revisited, using modern molecular structure codes to recompute the interaction potential to higher accuracy.

Excitation of C $2p^2\ ^3P$ and O $2p^4\ ^3P$ in collisions with H

Rate coefficients for fine structure transitions in C^0 and O^0 were computed by Launay and Roueff [128]. The uncertainties in these results are at least as large as for C^+–H^0 scattering, considered above. The applicability of the orbiting approximation is debatable for collisions between *neutral* particles, where the long-range polarization potential ($\sim R^{-4}$) is absent, and the leading term in the potential ($\sim R^{-6}$) is of shorter range. Once again, these calculations need to be repeated, using updated interaction potentials.

6.2 Theory of fine structure excitation processes

Table 6.5. *Coefficients of the fit (6.56) to the rate coefficient, $L(T)$, for cooling through collisional excitation of fine structure transitions by H^0, in units of 10^{-24} erg cm^3 s^{-1}. Original data for C^+, C^0 and O^0 from [127], [128] and for Si^+ from [132]; parameters for Fe^+ from [133].*

	a	α (K)	b	β (K)	γ
C^+	22.5	92	0.0	0	0.0
C^0	1.18	23	3.63	62	0.10
O^0	0.080	228	0.027	326	0.66
Si^+	14.6	413	0.0	0	0.28
Fe^+	110	554	154	961	0.0

6.2.3 Cooling rates

One of the main uses of the cross-sections for fine structure transitions is to determine the contribution of collisional excitation, followed by radiative decay, to the rate of cooling of the interstellar gas. At low densities, when collisional excitation is followed by radiative decay, the rate of cooling per unit volume through collisional excitation of atom (or ion) A by the atomic (or molecular) perturber P is given by $n(A)n(P)L(T)$, where T is the kinetic temperature of the gas and n denotes a number density. If the rate of cooling per unit volume is expressed in erg cm^{-3} s^{-1} and the densities are in units of cm^{-3}, then $L(T)$ is in erg cm^3 s^{-1}. [The SI unit of power (W) is 1 J s^{-1} ≡ 10^7 erg s^{-1}.]

A useful fit to the cooling rate coefficients, $L(T)$, in the low-density limit is provided by the formula

$$L(T) = [a \exp(-\alpha/T) + b \exp(-\beta/T)]T^\gamma \quad (6.56)$$

In Table 6.5 are listed values of the parameters a, b, α, β and γ, which reproduce the original data for excitation by H^0 – the principal perturber in gas of sufficiently low density – to within about 30%.

Roueff [132] calculated the rate of cooling through excitation of the $j = \frac{1}{2} \to \frac{3}{2}$ transition of Si^+ by H^0, in the limit of low density. Her results can be accurately fitted by

$$L_{Si^+}(T) = 14.6 \times 10^{-24} T^{0.28} \exp(-413/T)$$

in units of erg cm^3 s^{-1}. For Fe^+–H^0, Dalgarno and McCray [133] suggested

$$L_{Fe^+}(T) = 1.1 \times 10^{-22}[\exp(-554/T) + 1.4 \exp(-961/T)]$$

in the same units. These results are also given in Table 6.5.

The rates of cooling through H_2 impact excitation of Si^+ and Fe^+ remain unknown. By analogy with the rates computed for C^+, the cooling rate coefficients for H_2 might be taken to be 50% of those for H^0 impact. Quantal calculations are needed but have not yet been carried out, and these estimates are uncertain by at least a factor of 2.

The excitation of fine structure transitions

In the discussion so far, it has been assumed that the low-density limit applies, in which case the rate of cooling per unit volume varies quadratically with the gas density. For example, the rate of cooling per unit volume through collisional excitation of C^+ by H^0 may be written

$$n(C^+)n(H^0)L(T) = n_H^2 \left[\frac{n(C^+)}{n_H} \frac{n(H^0)}{n_H} L(T) \right]$$

where $n_H = n(H) + 2n(H_2) + n(H^+) + \cdots$ is the total density of hydrogen nuclei. When collisional excitation is followed by radiative decay back to the ground state, which is the case at low densities, and when the lines are optically thin, which is the case of transitions forbidden to electric dipole radiation,

$$L(T) = \sum_j \langle \sigma(j \leftarrow j')v \rangle \Delta E(j,j') \tag{6.57}$$

where j' is the lowest fine structure state, and

$$\langle \sigma(j \leftarrow j')v \rangle = \int \sigma(j \leftarrow j')v f(v,T) dv \tag{6.58}$$

In (6.58), $f(v,T)$ denotes the velocity distribution, taken to be Maxwellian [equation (5.30)]. The excitation energy of state j is $\Delta E(j,j')$. It follows from the principle of detailed balance that

$$\langle \sigma(j \leftarrow j')v \rangle (2j'+1) = \langle \sigma(j' \leftarrow j)v \rangle (2j+1) \exp\left[-\frac{\Delta E(j,j')}{k_B T} \right] \tag{6.59}$$

At high densities, on the other hand, equilibrium between collisional excitation and de-excitation prevails and

$$\frac{n_j}{n_{j'}} = \frac{2j+1}{2j'+1} \exp\left[-\frac{\Delta E(j,j')}{k_B T} \right] \tag{6.60}$$

that is, a Boltzmann distribution is established. In this case, the rate of cooling per unit volume is proportional to the number density of the atom or ion emitting the fine structure transition and increases linearly with the gas density.

Between the low and the high density limits lies the regime in which the level populations have to be evaluated from the equations of statistical equilibrium, allowing for collisional excitation and de-excitation and for radiative decay. This regime may be characterized by the critical value of the perturber density, n_{crit}, at which the rates of collisional and radiative de-excitation of a given fine structure state are equal. Low density formulae apply for $n(P) \ll n_{crit}$, and a Boltzmann distribution for $n(P) \gg n_{crit}$, the value of n_{crit} being dependent on the kinetic temperature, T. By way of illustration, the critical densities for fine structure transitions of C^0, C^+ and O^0 are listed in Table 6.6. These values of n_{crit} were calculated using the spontaneous radiative transition probabilities in Table 6.1 and the collisional rate coefficients for $P \equiv H^0$ [127, 128] for $T = 100$ K.

6.2 Theory of fine structure excitation processes

Table 6.6. *Critical densities, $n_{\mathrm{crit}}(\mathrm{H}^0)$, at which the rates of radiative de-excitation and collisional de-excitation by H^0 of the upper fine structure levels are equal, at $T = 100$ K. Numbers in parentheses are powers of 10.*

Atom/ion	Transition	n_{crit} (cm^{-3})
C^0	$^3P_1 \rightarrow {}^3P_0$	500
	$^3P_2 \rightarrow {}^3P_1$	700
C^+	$^2P^o_{\frac{3}{2}} \rightarrow {}^2P^o_{\frac{1}{2}}$	3100
O^0	$^3P_1 \rightarrow {}^3P_2$	9.8 (5)
	$^3P_0 \rightarrow {}^3P_1$	1.1 (5)

Table 6.7. *Sources of atomic data relating to fine structure transitions induced by ortho- and para-H_2 and by He.*

	H_2	He
C^+	[126]	
C^0	[134]	[135]; [136]
O^0	[137]	[138]

The low density formulae should apply in diffuse clouds, where most of the hydrogen is in atomic form. In dense clouds, on the other hand, most of the hydrogen is molecular, and the ortho : para composition is uncertain. Under these circumstances, the statistical equilibrium equations should be solved, allowing for collisions induced by molecular hydrogen and also by He. The sources of the requisite data are specified in Table 6.7. The more recent data are more secure, having been calculated using more accurate interaction potentials.

7

Radiative transfer in molecular lines

7.1 Introduction

Conditions of thermodynamic equilibrium are the exception, rather than the rule, in the interstellar medium. In order to interpret the observed intensities of molecular emission lines, it is usually necessary to know the relevant collisional and radiative transition rates. If the lines are optically thin, they do not undergo significant reabsorption within the region emitting the radiation, and the emitted flux is obtained as the line-of-sight integral of the rate of emission per unit volume of gas. However, it is often the case that strong emission lines are optically thick or, at least, have a significant optical depth (i.e. an optical depth of the order of 1) at their centres. Under these circumstances, it is necessary to solve the equation of radiative transfer in order to predict the emitted line fluxes to a reasonable degree of accuracy.

Solving radiative line transfer problems is no mean task. Both analytical and stochastic (Monte-Carlo) approaches are followed, with the latter being more readily applicable when the geometry or the density distribution does not admit simple treatments; this is likely to be always the case of interstellar molecular clouds. Unfortunately, it is also the case that the geometry and the density distribution are generally poorly known or unknown. Accordingly, treatments of the radiative transfer problem that go beyond simple approximations often lack the requisite observational constraints on input parameters. Usually, molecular line transfer problems are solved by means of the *large velocity gradient* approximation, discussed in the following section. This method is based on the assumption that the local velocity gradient at any point is sufficiently large to move photons into the optically thin wings of a line, over a distance that is small compared with the dimensions which characterize variations in the physical and chemical parameters of the cloud. The velocity gradient in molecular clouds arises from macroscopic motions, including (and perhaps principally) turbulence.

One of the most remarkable manifestations of departures from thermodynamic equilibrium in the interstellar medium is provided by observations of masers; the requirements for maser action are considered in the following section. Several molecules are known to have transitions that can mase, including OH, H_2O, SiO and methanol. Indeed, methanol is both a maser and an 'anti-maser' in different conditions and transitions, that is, it has levels that can become over-populated, relative to a Boltzmann distribution at the kinetic temperature of the gas, and a level that can be under-populated relative to a Boltzmann distribution at the temperature of the cosmic background radiation field. Maser spots have very small angular size and provide information on the physical conditions in well localized regions of interstellar space.

7.2 The radiative transfer equation

The processes leading to population inversion may be collisional or radiative. In many cases, the primary process has been shown to be radiative, generally associated with locally intense sources of infrared radiation. Even then, collisional processes are important in redistributing population and have to be taken into account in any quantitative model of the maser emission. Furthermore, there is at least one example, that of the 12.18 GHz $J_K = 2_0-3_{-1}$ transition of E-type methanol, where collisions must be responsible for the population anomaly, in this case by successfully competing with thermalization at the temperature (2.73 K) of the cosmic microwave background radiation [139]. Quantitative treatments of all these problems require that the equation of radiative transfer be solved. We shall consider first the radiative transfer equation, and then the first interstellar maser to be discovered, the OH radical, which has been extensively observed since its discovery.

7.2 The radiative transfer equation

Consider an atom or a molecule with energy levels $1, 2, 3, \ldots$, such that $E_1 < E_2 < E_3 < \ldots$, and let the population densities be $n_1, n_2, n_3, \ldots$. In thermal equilibrium at temperature T, the relative populations of the levels (e.g. 1 and 2) are determined by a Boltzmann distribution,

$$\frac{n_2}{n_1} = \frac{\omega_2}{\omega_1} \exp\left[-\frac{(E_2 - E_1)}{k_B T}\right] \tag{7.1}$$

where ω denotes a statistical weight (degeneracy) and k_B is Boltzmann's constant. Equation (7.1) may be written in the form

$$\frac{n_2/\omega_2}{n_1/\omega_1} = \exp\left[-\frac{(E_2 - E_1)}{k_B T}\right]$$

where n/ω denotes a population density per degenerate sub-state (ω is the number of states with the same energy E).

The rate of spontaneous radiative transitions from level 2 to level 1 per unit volume is given by $A(1 \leftarrow 2)n_2$, where A is the probability per unit time of a spontaneous transition. The corresponding rate of induced radiative transitions per unit volume depends not only on n_2 but also on the radiation density u_ν at the frequency ν of the line. The rate of induced radiative transitions is $u_\nu B(1 \leftarrow 2)n_2$, where the constant of proportionality is the Einstein B-coefficient. Similarly, the rate of stimulated upwards transitions is $u_\nu B(2 \leftarrow 1)n_1$. In equilibrium,

$$A(1 \leftarrow 2)n_2 + u_\nu B(1 \leftarrow 2)n_2 = u_\nu B(2 \leftarrow 1)n_1 \tag{7.2}$$

whence

$$u_\nu = \frac{A(1 \leftarrow 2)n_2}{B(2 \leftarrow 1)n_1 - B(1 \leftarrow 2)n_2} \tag{7.3}$$

Substituting the Boltzmann distribution (7.1), we obtain

$$u_\nu = \frac{A(1 \leftarrow 2)}{B(1 \leftarrow 2)} \frac{1}{\frac{B(2 \leftarrow 1)\omega_1}{B(1 \leftarrow 2)\omega_2} \exp[(E_2 - E_1)/(k_B T)] - 1} \tag{7.4}$$

In equilibrium, the radiation density is given by a Planck distribution at temperature T, namely

$$u_\nu = \frac{8\pi h \nu^3}{c^3} \frac{1}{\exp[(h\nu)/(k_B T)] - 1} \quad (7.5)$$

and where, in the example being considered, $h\nu = E_2 - E_1$. As equations (7.4) and (7.5) must be identical for all values of T, we obtain the familiar relationships between the Einstein A- and B-coefficients:

$$B(2 \leftarrow 1)\omega_1 = B(1 \leftarrow 2)\omega_2 \quad (7.6)$$

and

$$B(1 \leftarrow 2) = A(1 \leftarrow 2)\frac{c^3}{8\pi h \nu^3} \quad (7.7)$$

Let us now introduce the line profile function, ϕ_ν, normalized such that $\int \phi_\nu d\nu = 1$. Line profiles are usually determined by random thermal and turbulent motions in the emitting gas, in which case a *Doppler* profile

$$\phi_\nu = \frac{1}{\nu_0}\left(\frac{\beta}{\pi}\right)^{\frac{1}{2}} \exp\left[-\beta\left(\frac{\nu - \nu_0}{\nu_0}\right)^2\right] \quad (7.8)$$

or, equivalently, a *Gaussian* profile

$$\phi_\nu = \frac{2}{\Delta\nu}\left(\frac{\ln 2}{\pi}\right)^{\frac{1}{2}} \exp\left[-4 \ln 2 \left(\frac{\nu - \nu_0}{\Delta\nu}\right)^2\right] \quad (7.9)$$

is applicable. In equation (7.8), ν_0 is the frequency at the line centre and $\beta = mc^2/(2k_B T_D)$, where m is the mass of the emitting atom or molecule and T_D is the Doppler temperature. In equation (7.9), $\Delta\nu$ is the full width of the line at half-maximum intensity. Clearly,

$$\Delta\nu = 2\nu_0 (\ln 2/\beta)^{\frac{1}{2}} \quad (7.10)$$

The rate of upwards transitions at frequency ν in a line is $u_\nu B(2 \leftarrow 1)n_1 \phi_\nu$ and of stimulated downwards transitions $u_\nu B(1 \leftarrow 2)n_2 \phi_\nu$. Stimulated emission may be considered to be negative absorption; it preserves both the direction of propagation and the state of polarization of the stimulating photon. Spontaneous radiative emission, on the other hand, is random in both direction and sense of polarization. The density u_ν and the intensity I_ν of radiation propagating in an element of solid angle $d\omega$ in any given direction are related through

$$u_\nu = \frac{1}{c} I_\nu(\omega) d\omega \quad (7.11)$$

A fraction $d\omega/(4\pi)$ of the spontaneously emitted photons travel in the given direction. Hence, the change in the intensity of the radiation propagating in the x-direction, say, over an element of distance dx is given by

$$\frac{dI_\nu}{dx} = -\frac{h\nu}{c}[B(2 \leftarrow 1)n_1 - B(1 \leftarrow 2)n_2]\phi_\nu I_\nu + \frac{h\nu}{4\pi} A(1 \leftarrow 2)n_2 \phi_\nu \quad (7.12)$$

7.2 The radiative transfer equation

where $h\nu$ is the photon energy. Using equation (7.6), we obtain

$$\frac{dI_\nu}{dx} = -\kappa_\nu I_\nu + j_\nu \tag{7.13}$$

where

$$\kappa_\nu = \frac{h\nu}{c}\left(\frac{n_1}{\omega_1} - \frac{n_2}{\omega_2}\right) B(1 \leftarrow 2)\omega_2 \phi_\nu \tag{7.14}$$

is the *opacity* at the frequency ν in the line and

$$j_\nu = \frac{h\nu}{4\pi} A(1 \leftarrow 2) n_2 \phi_\nu \tag{7.15}$$

is the *emission coefficient*. Dividing the radiative transfer equation (7.13) by the opacity, κ_ν, we obtain

$$\frac{dI_\nu}{d\tau_\nu} = -I_\nu + S_\nu \tag{7.16}$$

where τ_ν is the *optical depth* at frequency ν in the line, and $S_\nu = j_\nu / \kappa_\nu$ is the *source function*. The optical depth is related to the opacity by

$$d\tau_\nu = \kappa_\nu dx \tag{7.17}$$

An alternative expression for the opacity is, using equation (7.7),

$$\kappa_\nu = \frac{c^2}{8\pi \nu^2}\left(\frac{n_1}{\omega_1} - \frac{n_2}{\omega_2}\right) A(1 \leftarrow 2)\omega_2 \phi_\nu \tag{7.18}$$

Using equations (7.15) and (7.18), the source function becomes

$$S_\nu = \frac{2h\nu^3}{c^2} \frac{1}{\frac{n_1/\omega_1}{n_2/\omega_2} - 1} \tag{7.19}$$

In thermodynamic equilibrium at temperature T, the ratio $(n_1/\omega_1)/(n_2/\omega_2)$ is determined by the Boltzmann relation, equation (7.1), and the source function, S_ν, becomes identical to the Planck function, B_ν, defined by

$$B_\nu = \frac{2h\nu^3}{c^2} \frac{1}{\exp[(h\nu)/(k_B T)] - 1} \tag{7.20}$$

In general, thermodynamic equilibrium does not apply, and the intensity at frequency ν in the line must be obtained by solving the radiative transfer equation (7.13) – a non-trivial task, even today.

7.2.1 The 'Large Velocity Gradient' approximation

In the large velocity gradient (LVG) model, the opacity κ_ν in a line is supposed to be limited by the Doppler shift arising from a locally large macroscopic velocity gradient in the medium. To a velocity shift, dv, there corresponds a Doppler shift in frequency, $d\nu$, such that

$$\frac{dv}{v} = \frac{d\nu}{c}$$

or

$$dv = \frac{c}{\nu} d\nu \quad (7.21)$$

The velocity of the gas is a function of position along the line of sight, i.e. $v = v(x)$. Writing

$$dx = \frac{dx}{dv} dv$$

and using equation (7.21), we obtain

$$dx = \frac{1}{(dv/dx)} \frac{c}{\nu} d\nu \quad (7.22)$$

Using equations (7.17), (7.18) and (7.22), the derivative of the optical depth, τ_ν, becomes

$$d\tau_\nu = \frac{c^3}{8\pi \nu^3} \left(\frac{n_1}{\omega_1} - \frac{n_2}{\omega_2} \right) A(1 \leftarrow 2) \omega_2 \phi_\nu \frac{1}{(dv/dx)} d\nu \quad (7.23)$$

If the line width is small compared with the central frequency of the line, ν_0, then the variation of the profile function, ϕ_ν, with ν much faster than ν^3, and so equation (7.23) may be integrated with respect to ν, yielding

$$\tau_{\nu_0} \approx \frac{c^3}{8\pi \nu_0^3} \left(\frac{n_1}{\omega_1} - \frac{n_2}{\omega_2} \right) A(1 \leftarrow 2) \omega_2 \frac{1}{(dv/dx)} \quad (7.24)$$

where we have used the normalization condition $\int \phi_\nu d\nu = 1$. Assuming that the level population densities and hence the source function, S_ν [equation (7.19)], are constant, the equation of radiative transfer (7.16) at the line centre, $\nu = \nu_0$, may be integrated, giving

$$\left[\ln(-I_{\nu_0} + S_{\nu_0}) \right]_{I_{\nu_0}(0)}^{I_{\nu_0}(x)} = -\left[\tau_{\nu_0} \right]_0^{\tau_{\nu_0}(x)} \quad (7.25)$$

If the background radiation field is a black body at temperature T_{bb}, $I_{\nu_0}(0) \equiv B_{\nu_0}(T_{bb})$, where B_{ν_0} is the Planck function at frequency ν_0. Then, at distance x into the cloud,

$$I_{\nu_0} = S_{\nu_0}(1 - e^{-\tau_{\nu_0}}) + B_{\nu_0}(T_{bb}) e^{-\tau_{\nu_0}} \quad (7.26)$$

The first term on the right-hand side of equation (7.26) is the contribution to the radiation intensity from photons produced within the cloud, whereas the second term is the attenuated intensity of the background radiation.

The intensity of the radiation that is emitted in the cloud over the optical path length τ_{ν_0} is

$$\int j_{\nu_0} dx \equiv \int S_{\nu_0} d\tau_{\nu_0} = S_{\nu_0} \tau_{\nu_0}$$

7.2 The radiative transfer equation

when S_ν is constant. Comparing the intensity of radiation which emerges [first term on the right-hand side of equation (7.26)] with the radiation which is produced within the source, we may define the *escape probability* of a photon produced within the source as

$$\beta = \frac{S_{\nu_0}(1 - e^{-\tau_{\nu_0}})}{S_{\nu_0}\tau_{\nu_0}} = \frac{(1 - e^{-\tau_{\nu_0}})}{\tau_{\nu_0}} \tag{7.27}$$

We see from equations (7.26) and (7.27) that, as $\tau_{\nu_0} \to 0$, $\beta \to 1 - 1/2\tau_{\nu_0}$ and $I_{\nu_0} \to S_{\nu_0}\tau_{\nu_0} + B_{\nu_0}$, whereas, as $\tau_{\nu_0} \to \infty$, $\beta \to \frac{1}{\tau_{\nu_0}}$ and $I_{\nu_0} \to S_{\nu_0}$. In the former, *optically thin* case, all the radiation that is produced within the source is emitted, and the background radiation is unattenuated. In the latter, *optically thick* case, radiation is reabsorbed at the point at which it is emitted, ('on the spot') and the radiation intensity becomes equal to the source function. The on-the-spot approximation has been used extensively to treat the ionizing radiation produced internally by gaseous nebulae.

The escape probability (7.27) may be incorporated into the equations that determine the populations of the molecular energy levels, in order to take account (approximately) of the finite optical depths in the lines. The rate (s^{-1}) of stimulated emission, owing to a background black-body radiation field and the internally produced radiation, is

$$\frac{4\pi}{c}[W\beta B_\nu + (1 - \beta)S_\nu]B(1 \leftarrow 2)$$

where S_ν and B_ν are given by equations (7.19) and (7.20); W is the geometrical dilution factor applicable to the external field. In the above expression, the factor $4\pi/c$ converts the radiation intensity to the radiation density. Using equation (7.7), the rate of stimulated emission becomes

$$\left[\frac{W\beta}{\exp[(h\nu_0)/(k_B T_{bb})] - 1} + \frac{1 - \beta}{((n_1/\omega_1)/(n_2/\omega_2)) - 1}\right]A(1 \leftarrow 2)$$

to which the spontaneous radiative transition probability, $A(1 \leftarrow 2)$, must be added to yield the total downwards ($1 \leftarrow 2$) radiative rate. Using (7.6), the rate (s^{-1}) of induced absorption is

$$\left[\frac{W\beta}{\exp[(h\nu_0)/(k_B T_{bb})] - 1} + \frac{1 - \beta}{((n_1/\omega_1)/(n_2/\omega_2)) - 1}\right]A(1 \leftarrow 2)\frac{\omega_2}{\omega_1}$$

Thus, the radiative rates of transfer of population between the levels 1 and 2 depend explicitly and implicitly (via the escape probability) on the level populations themselves. The level populations and the escape probability have to be calculated self-consistently, usually by means of an iterative method.

7.2.2 Population inversion and maser action

The terms 'laser' and 'maser' are acronyms whose common letters represent **a**mplification by **s**timulated **e**mission of **r**adiation. We have already noted that spontaneous radiative emission is random in both direction and sense of polarization, and hence it does not intervene in the maser mechanism. Accordingly, the emission coefficient, j_ν [equation (7.15)], may be dropped from the equation of radiative transfer (7.13). Then, if κ_ν is independent of position, x, the transfer equation can be integrated, yielding

$$I_\nu(x) = I_\nu(0)\exp(-\kappa_\nu x) \tag{7.28}$$

It is customary, when discussing relative level populations in astronomical sources, to introduce the *excitation temperature*, T_{ex}, defined by analogy with equation (7.1) such that

$$\frac{n_2/\omega_2}{n_1/\omega_1} = \exp[-(E_2 - E_1)/(k_B T_{\text{ex}})] \tag{7.29}$$

Recalling that $(E_2 - E_1) = h\nu$, the excitation temperature may be introduced into the equation (7.18) for the opacity, giving

$$\begin{aligned}\kappa_\nu &= \frac{c^2}{8\pi\nu^2}\left[1 - \exp\left(-\frac{h\nu}{k_B T_{\text{ex}}}\right)\right]\frac{n_1}{\omega_1}A(1 \leftarrow 2)\omega_2\phi_\nu \\ &\approx \frac{c^2}{8\pi\nu^2}\frac{h\nu}{k_B T_{\text{ex}}}\frac{n_1}{\omega_1}A(1 \leftarrow 2)\omega_2\phi_\nu\end{aligned} \tag{7.30}$$

where the approximate expression applies if $h\nu \ll k_B T_{\text{ex}}$.

It may be seen from equations (7.29) and (7.30) that, when $n_2/\omega_2 < n_1/\omega_1$, which is the case in thermodynamic equilibrium, $T_{\text{ex}} > 0$ and $\kappa_\nu > 0$; then the radiation is attenuated on passing through the gas. On the other hand, if $n_2/\omega_2 > n_1/\omega_1$, $T_{\text{ex}} < 0$ and $\kappa_\nu < 0$, resulting in an exponential increase in the radiation intensity (maser action).

Another parameter that is used extensively in observational radio astronomy is the *brightness temperature*, T_b, defined as the temperature of a black-body of the same intensity at the same frequency,

$$I_\nu = B_\nu(T_b) \tag{7.31}$$

where B_ν is the Planck function [equation (7.20)]. At radio frequencies, $h\nu/k_B T \ll 1$ and so

$$B_\nu \approx 2\left(\frac{\nu}{c}\right)^2 k_B T \tag{7.32}$$

The condition $n_2/\omega_2 > n_1/\omega_1$ is known as 'population inversion'. If this condition can be established over a sufficient path length, high radiation brightness temperatures are produced by amplification of the incident radiation field, $I_\nu(0)$, which is produced by a background source (and might be, for example, the thermal emission from an H II region).

7.3 The OH radical

The OH radical was the first interstellar 'molecule' to be detected with the techniques of radio astronomy [140]. Prior to this discovery, a few interstellar radicals (CH, CH$^+$ and CN) had been observed through their optical emission line spectra. Our knowledge of interstellar molecules, and of the media in which they exist, subsequently increased rapidly with the development of radiofrequency, and particularly millimetric, receivers. We shall take OH as a template for interstellar masers and discuss the role of collision processes in establishing its level populations; but first we need to consider the internal structure of the OH radical.

7.3 The OH radical

The OH transitions observed by Weinreb *et al.* [140] were those between the components of the ground state (X$^2\Pi_{\frac{3}{2}} J = \frac{3}{2}$) Λ-doublet. The OH radical has an unpaired electron in its ground electronic state, in which the projection of electronic orbital angular momentum along the internuclear axis is $\Lambda = 1$. The resultant electronic angular momentum is

$$\Omega = \Lambda + \Sigma$$

where $\Sigma = \frac{1}{2}$ is the projection of the spin angular momentum on the internuclear axis, and $\Omega = \frac{1}{2}$ or $\Omega = \frac{3}{2}$ is the resultant electronic angular momentum. The further coupling to the nuclear rotational angular momentum, **R**, of the radical

$$\mathbf{J} = \mathbf{\Omega} + \mathbf{R}$$

gives rise to rotational ladders corresponding to each of the electronic states X$^2\Pi_{\frac{1}{2}}$ and X$^2\Pi_{\frac{3}{2}}$, where the subscript denotes the value of Ω. The lowest level in each ladder is $J = \Omega$.

The above angular momentum coupling scheme, which supposes that both the electronic orbital and spin angular momenta are strongly bound to the internuclear axis, is known as 'Hund's case a'. The Hund's cases are discussed, for example, by Herzberg [120, 141]; they are, in essence, different angular momentum coupling schemes that sometimes provide good approximations to those occurring in molecules. In case b (which was adopted implicitly when discussing the formation of the H$_2$ molecule in Section 1.2.1), the spin vector, **S** is supposed to be weakly coupled to the internuclear axis and to **Λ**, and the total angular momentum is composed from

$$\mathbf{N} = \mathbf{\Lambda} + \mathbf{R}$$

and

$$\mathbf{J} = \mathbf{N} + \mathbf{S}$$

As the nuclear rotational angular momentum **R** increases, the 'case b' limit is approached. However, the low rotational states J of OH are intermediate between case a and case b, and calculations should be carried out in intermediate coupling. Then, the wave functions for given J^p, where p is the parity, can be expressed as linear combinations of the case a eigenfunctions. The correctly symmetrized and normalized case a eigenfunctions may be written [142, 143]

$$|JM_J\Omega\epsilon\rangle = 2^{-\frac{1}{2}}(|JM_J\Omega\rangle|\Lambda\Sigma\rangle + \epsilon|JM_J,-\Omega\rangle|-\Lambda,-\Sigma\rangle) \tag{7.33}$$

where the electronic component of the case a eigenfunction is denoted $|\Lambda\Sigma\rangle$, and the rotational component is

$$|JM_J\Omega\rangle = \left(\frac{2J+1}{8\pi^2}\right)^{\frac{1}{2}} D^{J*}_{M_J\Omega}(\alpha,\beta,\gamma) \tag{7.34}$$

where **D** is the rotation matrix; (α, β, γ) are the Euler angles [48]. The projection of **J** on the space-fixed z-axis is denoted M_J, whilst its projection on the body-fixed Z-axis (the

Table 7.1. *Numerical values of the coefficients a_J and b_J which determine the degree of mixing of rotational states J in the $X^2\Pi_{\frac{1}{2}}$ and $X^2\Pi_{\frac{3}{2}}$ rotational ladders of OH in its ground electronic state; see equation (7.35).*

J	a_J	b_J
1/2	1	0
3/2	0.9848	0.1739
5/2	0.9642	0.2653
7/2	0.9421	0.3354
9/2	0.9209	0.3898
11/2	0.9017	0.4324
∞	0.7071	0.7071

internuclear axis of the radical) is Ω. Rotations through the Euler angles (α, β, γ) carry the space-fixed into the body-fixed coordinate system. The parity of the eigenfunctions (7.33) is $p = (-1)^{J-S}\epsilon$ [144–147]. In intermediate coupling (in this case, coupling that is intermediate between cases a and b), Ω is no longer a good quantum number, and states with the same values of J and p in the two ladders, $\Omega = \frac{1}{2}$ and $\Omega = \frac{3}{2}$, are 'mixed':

$$\psi_{\frac{3}{2}}(J^p) = a_J \psi_{\frac{3}{2}}^{(a)}(J^p) + b_J \psi_{\frac{1}{2}}^{(a)}(J^p)$$
$$\psi_{\frac{1}{2}}(J^p) = -b_J \psi_{\frac{3}{2}}^{(a)}(J^p) + a_J \psi_{\frac{1}{2}}^{(a)}(J^p) \quad (7.35)$$

where the superscript (a) denotes a case a eigenfunction. In (7.35),

$$b_J = \left[\frac{X - 2 + \lambda}{2X}\right]^{\frac{1}{2}}$$
$$a_J^2 + b_J^2 = 1 \quad (7.36)$$

and

$$X = [(2J+1)^2 + \lambda(\lambda - 4)]^{\frac{1}{2}} \quad (7.37)$$

The parameter $\lambda = A/B$ is the ratio of the spin-orbit interaction constant A and the rotational constant B of the radical; $|\lambda| \to \infty$ in the Hund's case a limit, whereas $\lambda \to 0$ in the case b limit. Values of the 'mixing coefficients', a_J and b_J, evaluated using the spectroscopic value of $\lambda = -7.501$ [148–150], are given in Table 7.1.

Figure 7.1 shows the three lowest rotational levels of OH in each of the $X^2\Pi_{\frac{1}{2}}$ and $X^2\Pi_{\frac{3}{2}}$ ladders, with the J^p and Ω^ϵ labelling. The so-called 'Λ-doubling' is the splitting that occurs between states with given values of Ω and J but different values of ϵ or p. In the lowest rotational state of OH, $J = \Omega = 3/2$, this splitting amounts to approximately 1/18 cm^{-1}. There is a further – and much smaller – splitting of the energy levels, not shown in Fig. 7.1,

7.3 The OH radical

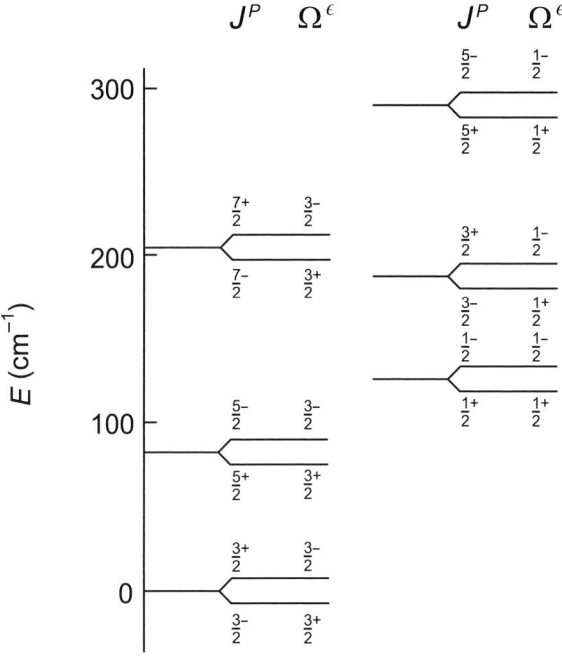

Figure 7.1 Illustrating the three lowest rotational levels of OH in each of the $X^2\Pi_{\frac{1}{2}}$ and $X^2\Pi_{\frac{3}{2}}$ ladders, with the corresponding J^p and Ω^ϵ labelling. The Λ-doubling, i.e. the splitting of levels of given Ω and J and different ϵ or p, is exaggerated for clarity of presentation.

owing to the interaction between **J** and the resultant nuclear spin, **I**. Denoting the total angular momentum by **F**, we have that

$$\mathbf{F} = \mathbf{J} + \mathbf{I}$$

and, as $J = 3/2$ in the ground state and $I = 1/2$ (owing to the unpaired spin of the proton), $F = 1$ or 2. These states are shown schematically in Fig. 7.2.

In an optically thin gas in thermodynamic equilibrium, the approximate relative intensities of the transitions, shown in Fig. 7.2, at 1667, 1665, 1612 and 1720 MHz are, respectively, 9 : 5 : 1 : 1. The 1667 and 1665 MHz transitions are called the 'main lines', and those at 1612 and 1720 MHz are called the 'satellite lines'. In an optically thick gas, all the lines should have equal intensity. 'Anomalous' relative line intensities were observed in W49 and NGC 6334 by Weaver et al. [151]. In particular, the intensity of the line at 1665 MHz was observed to exceed that of the 1667 MHz line, a situation which cannot prevail in thermodynamic equilibrium. The assumption of thermodynamic equilibrium must be abandoned in order to understand the observed line intensities; but at the time that the observations were first made, it was suggested that blending of the 1665 MHz transition with an unidentified line was responsible for the anomaly.

Subsequent observations have shown that the OH lines have brightness temperatures that exceed by many orders of magnitude the kinetic temperature of the emitting gas

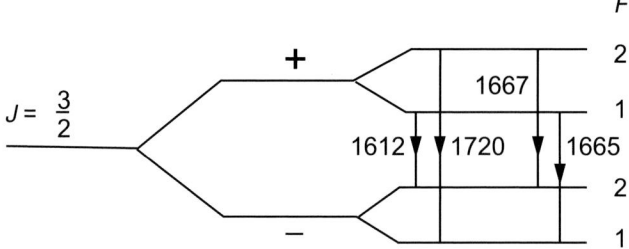

Figure 7.2 The $X^2\Pi_{\frac{3}{2}} J = \frac{3}{2}$ ground rotational state of OH, showing both the Λ-doubling, the parity ($\pm$), and the frequencies (in MHz) of the allowed transitions between the hyperfine states, F.

and the ambient radiation temperature. Furthermore, the lines are highly polarized. These characteristics establish beyond doubt that population inversion is occurring, giving rise to maser emission.

7.4 Producing population inversion

The hydroxyl radical (OH) is a good template to adopt when discussing interstellar masers, not only because it was the first maser to be observed in the interstellar medium, but also because it has remained an important means of studying the interstellar gas, principally in association with star formation. Furthermore, a whole range of mechanisms has been proposed to account for population inversion in OH, including pumping by radiation, in the ultraviolet and in the infrared, and both chemical and collisional pumping. While it seems certain now that infrared pumping is the primary mechanism leading to population inversion, the rates of collisional population transfer still need to be known, as collisions modify and can even quench maser action.

Ultraviolet radiative pumping was proposed by Litvak [152]; but this mechanism is highly energy inefficient and imposes severe constraints on the ultraviolet energy source. Furthermore, if ultraviolet photons are present in the gas, photodissociation of the OH radical can also take place. Gwinn *et al.* [153] considered the excitation of OH through the collisional dissociation of H_2O by H, in the reactions

$$H_2O + H \rightarrow OH + H_2$$
$$H_2O + H \rightarrow OH + H + H$$

which are endoergic by 0.64 and 5.12 eV, respectively; they considered also direct excitation in collisions of OH with H and H_2. The collisional excitation of OH by H, H_2, or He has been discussed subsequently by a number of authors [146, 147, 154–162]. Collisional excitation by charged particles was considered in [163–167]. Non-thermal excitation of OH by beams of charged particles, through the process of *ambipolar diffusion* (see Chapter 2) was discussed in [168, 169]. However, ambipolar diffusion gives rise to heating of the gas, which leads to the removal of OH [170] in the reaction

$$OH + H_2 \rightarrow H_2O + H$$

which has a barrier of 1490 K (0.13 eV).

7.5 Rotational excitation of OH by H_2

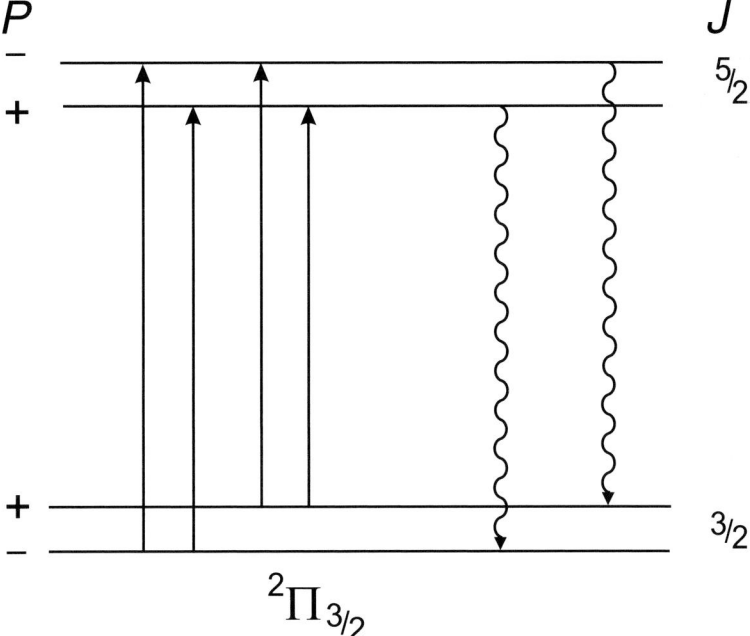

Figure 7.3 Producing anomalous populations in the ground state Λ-doublet of OH: collisional and radiative transitions between the $X^2\Pi_{\frac{3}{2}} J = \frac{3}{2}$ and $J = \frac{5}{2}$ Λ-doublets.

The rotational excitation of OH by any of the principal perturbers, H, H_2 and He, followed by radiative decay, could possibly lead to inversion of the populations of the levels of the ground state Λ-doublet. This mechanism relies on there being preferential excitation of one component of an excited state Λ-doublet (see the discussion in the following section). Subsequent radiative decay (according to electric dipole selection rules: $\Delta J = 0, \pm 1$; change of parity) then leads to over-population (relative to thermodynamic equilibrium) of one component of the ground state Λ-doublet (see Fig. 7.3).

Radiative pumping of the OH 18 cm maser through infrared rotational transitions was proposed by Litvak [171] and subsequently studied in more detail by Lucas [172] and Guilloteau *et al.* [173]. This mechanism relies on overlaps of the infrared transitions leading to population transfer among the hyperfine levels of the ground state Λ-doublet and hence to population inversion. The overlaps arise through the Doppler effect, associated with either the relative motion of different parts of the same gas cloud or thermal and microturbulent motions of the gas.

7.5 Rotational excitation of OH by H_2

The magnetic field present in the medium gives rise to Zeeman splitting of the magnetic sub-states, M_F, of the hyperfine states, F. Magnetic field strengths of the order of 10 mG have been deduced from observations of Zeeman pairs in the sources such as W3(OH) [174, 175]. The modelling of maser action involves solving the equation of radiative transfer in the presence of Zeeman splitting, allowing for population transfer through both radiative and collisional processes. Gray [176] has summarized and compared current models of polarized maser emission. Here we consider the transfer of population through collisional processes. Of the possible collision partners of OH, the H_2 molecule is likely to be the most important, as OH masers occur in molecular gas. Indeed, the formation of the OH radical, whether in shocks or in the cold ambient gas, requires that H_2 should be already present in the medium.

7.5.1 The OH–H_2 interaction potential

The first, semi-quantitative evaluation of the OH–H_2 interaction potential was performed by Bertojo *et al.* [154]. They made a number of simplifications, including the explicit treatment of the H_2 molecule as a system of spherical symmetry: the H_2 molecule was represented as a positive core with two 1s electrons, which is analogous to the ground state of a He atom. The results of such calculations are relevant to studies of collisions with para-H_2 molecules, constrained to their $J = 0$ ground rotational state, but not to collisions with ortho-H_2, for which $J > 0$.

The OH radical is chemically reactive because of incomplete pairing of its valence electrons. Of the three $2p\pi$ electrons, two are 'paired' and have anti-parallel spins, whereas the third is unpaired and determines the net electronic angular momenta of the radical. [We recall that 'p' denotes an orbital angular momentum $l = 1$, and π denotes an $m_l = \pm 1$ molecular state.] As we shall see below, the two paired electrons have a probability distribution such that they tend to lie in a plane that is perpendicular to the OH internuclear axis. The probability distribution of the unpaired electron is also perpendicular to the internuclear axis, and to the paired electrons; see Fig. 7.4.

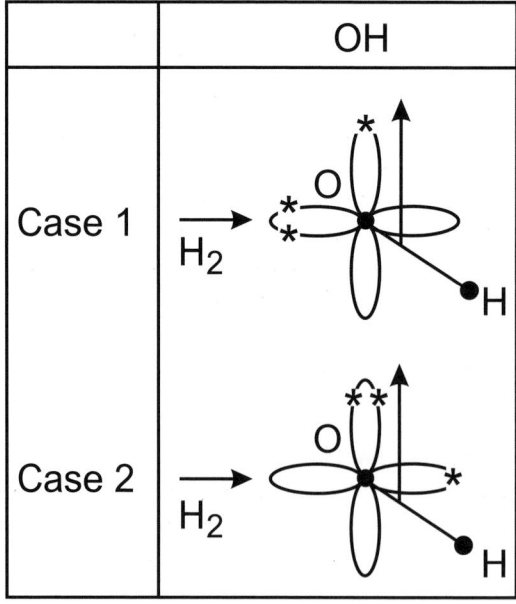

Figure 7.4 Illustrating the electron lobes of OH and the 'case 1' and 'case 2' collisions [154].

7.5 Rotational excitation of OH by H_2

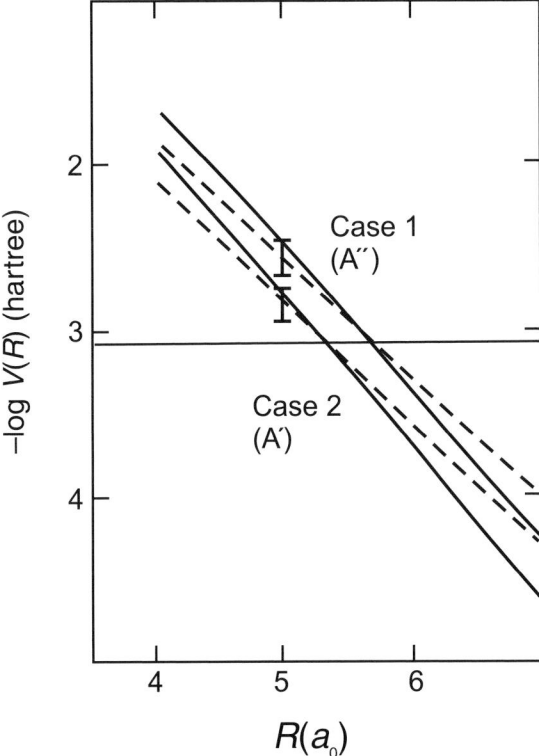

Figure 7.5 Case 1 (A″) and case 2 (A′) SCF interaction energies as functions of the separation of the OH and H_2 molecules. Dashed curves and error bars: Bertojo *et al.* [154]; continuous curves: Kochanski and Flower [177].

Bertojo *et al.* carried out calculations of the OH–H_2 interaction potential by means of the Hartree–Fock self consistent field (SCF) approximation, for separations of the molecular centres of mass $4 \leq R \leq 7a_0$, with the intermolecular vector **R** perpendicular to the O–H internuclear axis. A distinction was made between 'case 1' collisions, in which the two *paired* $2p\pi$ electrons have a high probability of being found in the H_2–O–H (collision) plane, and 'case 2' collisions, in which this probability is high for the *unpaired* $2p\pi$ electron. These cases are sketched in Fig. 7.4. As the H_2 molecule contains two paired electrons, with antiparallel spins, the electronic interaction in case 1 should be more repulsive than in case 2; this expectation was confirmed by their computations of the interaction energies (see Fig. 7.5).

The qualitative behaviour of the potential energy curves was confirmed by subsequent and more complete calculations of the OH–H_2 potential energy surfaces [177]. In the language of group theory, 'case 1' and 'case 2' correspond to the irreducible representations A″ and A′ of the C_s point group, in which the collision plane is a plane of symmetry of the system (cf. the discussion of the excitation of the fine structure transition of C^+ by H_2 in the sub-section entitled 'Excitation of C^+ $2p\,^2P^o$ in collisions with H_2', in Section 6.2.1).

We have mentioned already that the OH radical has one unpaired $2p\pi$ electron which determines the essential electronic properties. The angular part of the wave function

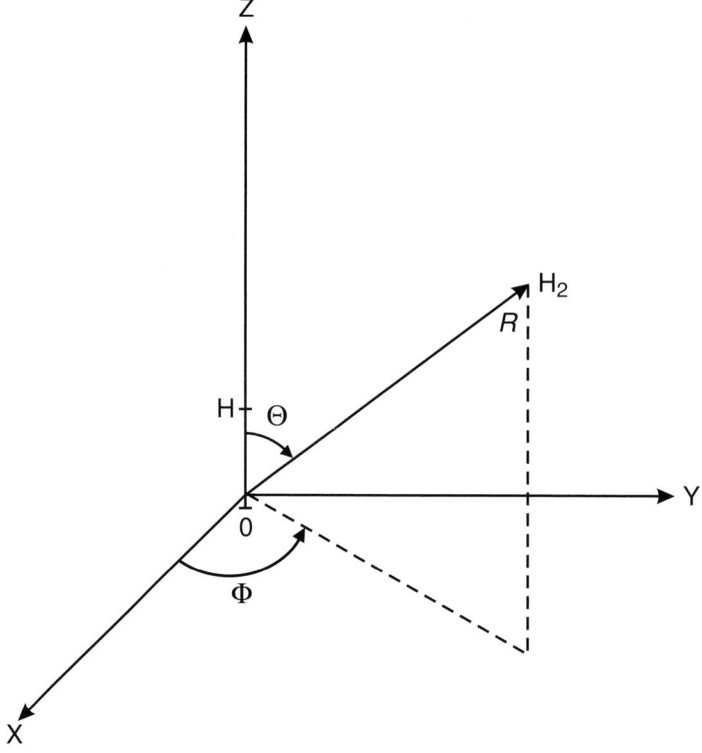

Figure 7.6 Illustrating the coordinates (R, Θ, Φ) of the centre of mass of the H_2 molecule, in the body-fixed frame.

representing such an electron may be denoted $Y_{1,\pm 1}(\hat{\mathbf{r}}'_1)$, where $\hat{\mathbf{r}}'_1 = (\theta'_1, \phi'_1)$ are the polar angles of the electron in the body-fixed frame, in which the Z-axis coincides with the OH internuclear axis. The function $Y_{1,\pm 1}$ is a spherical harmonic, and the indices indicate that a p ($l = 1$) electron is being considered in a π ($m_l = \pm 1$) molecular state.

Let the YZ-plane of the body-fixed coordinate system (Fig. 7.6) be a plane of symmetry of the radical. The electronic eigenfunction has a well-defined symmetry under reflection in this plane; the corresponding reflection operator will be denoted σ_{YZ}. The spherical harmonics $Y_{1,\pm 1}(\hat{\mathbf{r}}'_1)$ are *not* eigenfunctions of σ_{YZ}, as

$$\sigma_{YZ} Y_{1,\pm 1}(\hat{\mathbf{r}}'_1) = -Y_{1,\mp 1}(\hat{\mathbf{r}}'_1) \tag{7.38}$$

On the other hand, the normalized linear combinations

$$\frac{[Y_{11}(\hat{\mathbf{r}}'_1) \pm Y_{1,-1}(\hat{\mathbf{r}}'_1)]}{2^{\frac{1}{2}}}$$

are eigenfunctions of σ_{YZ}, with the eigenvalues ∓ 1. Recalling that

$$Y_{11}(\hat{\mathbf{r}}'_1) = -\left[\frac{3}{8\pi}\right]^{\frac{1}{2}} \sin \theta_1 e^{i\phi_1} \tag{7.39}$$

7.5 Rotational excitation of OH by H_2

and

$$Y_{1,-1}(\hat{\mathbf{r}}'_1) = \left[\frac{3}{8\pi}\right]^{\frac{1}{2}} \sin\theta_1 e^{-i\phi_1} \tag{7.40}$$

it may be seen that

$$Y_{11}(\hat{\mathbf{r}}'_1) + Y_{1,-1}(\hat{\mathbf{r}}'_1) \propto \sin\theta_1 \sin\phi_1$$

and

$$Y_{11}(\hat{\mathbf{r}}'_1) - Y_{1,-1}(\hat{\mathbf{r}}'_1) \propto \sin\theta_1 \cos\phi_1$$

corresponding to the lobes of the electronic wave function being along the Y-axis ($\theta_1 = \pi/2 = \phi_1$) or along the X-axis ($\theta_1 = \pi/2, \phi_1 = 0$), respectively. In the former case, the lobes of the angular distribution of the two paired electrons lie along the X-axis, and along the Y-axis in the latter case.

Suppose now that the H_2 molecule is incident in the YZ-plane, along the Y-axis. Because H_2 has a closed-shell electronic structure, its interaction with the two paired electrons of OH is more repulsive than its interaction with the unpaired electron. It follows that the OH–H_2 interaction is more repulsive in the electronic state

$$\frac{[Y_{11}(\hat{\mathbf{r}}'_1) - Y_{1,-1}(\hat{\mathbf{r}}'_1)]}{2^{\frac{1}{2}}}$$

than in the electronic state

$$\frac{[Y_{11}(\hat{\mathbf{r}}'_1) + Y_{1,-1}(\hat{\mathbf{r}}'_1)]}{2^{\frac{1}{2}}}$$

The former corresponds to A″ and the latter to A′ irreducible representations of the C_v point group.

The OH–para-H_2 interaction potential may be expanded as

$$V(\mathbf{R}) = (4\pi)^{\frac{1}{2}} \sum_{\lambda\mu} V_{\lambda\mu}(R) Y_{\lambda\mu}(\hat{\mathbf{R}}) \tag{7.41}$$

where $\mathbf{R} = (R, \Theta, \Phi)$ are the polar coordinates of the centre of mass of the (para-H_2) molecule in the body-fixed coordinate frame (see Fig. 7.6). When referring to 'para-H_2' in this context, we understand that the molecule is restricted to its rotational ground state, $J = 0$, and hence may be treated as a spherically symmetric perturber. The extension to the more general case of $H_2(J > 0)$ will be considered below. Following Alexander [142] and Dewangan et al. [143], the potential $V(\mathbf{R})$ may be written explicitly in terms of the A′ and A″ potential energy surfaces, in the form

$$V(R, \Theta, \Phi) = \frac{V_{A''}(R, \Theta) + V_{A'}(R, \Theta)}{2}$$
$$+ \frac{V_{A''}(R, \Theta) - V_{A'}(R, \Theta)}{2}(e^{2i\Phi} + e^{-2i\Phi}) \tag{7.42}$$

When the *ab initio* calculations of the A′ and A″ surfaces are suitably averaged with respect to the orientations of the of the H$_2$ internuclear axis, the resulting interaction energies, $V_{A''}$ and $V_{A'}$, may be used to derive the coefficients of the expansion (7.41), $V_{\lambda\mu}$, with $(\lambda,\mu) = (0,0), (1,0), (2,0), (2,2)$ and $(2,-2)$ [$V_{2,-2} = V_{2,2}$]. The explicit relations are

$$V(R,0,0) = V_{A''}(R,0) = V_{A'}(R,0)$$
$$= V_{0,0}(R) + 3^{\frac{1}{2}} V_{1,0}(R) + 5^{\frac{1}{2}} V_{2,0}(R) \tag{7.43}$$

$$V(R,\pi,0) = V_{A''}(R,\pi) = V_{A'}(R,\pi)$$
$$= V_{0,0}(R) - 3^{\frac{1}{2}} V_{1,0}(R) + 5^{\frac{1}{2}} V_{2,0}(R) \tag{7.44}$$

$$V(R,\pi/2,0) = \frac{3}{2} V_{A''}(R,\pi/2) - \frac{1}{2} V_{A'}(R,\pi/2)$$
$$= V_{0,0}(R) - \frac{5^{\frac{1}{2}}}{2} V_{2,0}(R) + \frac{30^{\frac{1}{2}}}{2} V_{2,2}(R) \tag{7.45}$$

and

$$V(R,\pi/2,\pi/2) = \frac{3}{2} V_{A'}(R,\pi/2) - \frac{1}{2} V_{A''}(R,\pi/2)$$
$$= V_{0,0}(R) - \frac{5^{\frac{1}{2}}}{2} V_{2,0}(R) - \frac{30^{\frac{1}{2}}}{2} V_{2,2}(R) \tag{7.46}$$

The coefficients $V_{\lambda\mu}(R)$, which were obtained from equations (7.43–7.46), are plotted in Fig. 7.7.

We now consider the generalization of equation (7.41) to the case of collisions with H$_2$ molecules in rotational states $J > 0$, notably ground state ortho-H$_2$, for which $J = 1$. This generalization involves introducing the coordinates which define the orientation of the H$_2$ internuclear axis.

The coordinate system to which we have been referring as the 'body-fixed frame' (cf. Fig. 7.6) is fixed in the OH molecule. A rotation through the Euler angles (Φ, Θ, Ψ) takes this molecule-fixed frame into the collision frame, in which the line joining the centre of mass of the OH radical to the centre of mass of the H$_2$ molecule (i.e. the intermolecular vector) is the z-axis; the third of the Euler angles, Ψ, is the rotation about the intermolecular vector and may be taken equal to 0. In the collision frame, the angular coordinates of the internuclear axis of the H$_2$ molecule are (θ_2, ϕ_2), and the interaction potential may be written as

$$V(R, \Theta, \Phi, \theta_2, \phi_2) = \sum_{\lambda_1 \mu \lambda_2 \nu} v_{\lambda_1 \mu \lambda_2 \nu}(R) D_{\mu\nu}^{\lambda_1 *}(\Phi, \Theta, 0) Y_{\lambda_2, -\nu}(\theta_2, \phi_2) \tag{7.47}$$

where $D_{\mu\nu}^{\lambda_1}(\Phi, \Theta, 0)$ is a rotation matrix element [48, 49] and $Y_{\lambda_2, -\nu}(\theta_2, \phi_2)$ is a spherical harmonic. The projections on the intermolecular axis (ν and $-\nu$) have opposite signs owing

7.5 Rotational excitation of OH by H_2

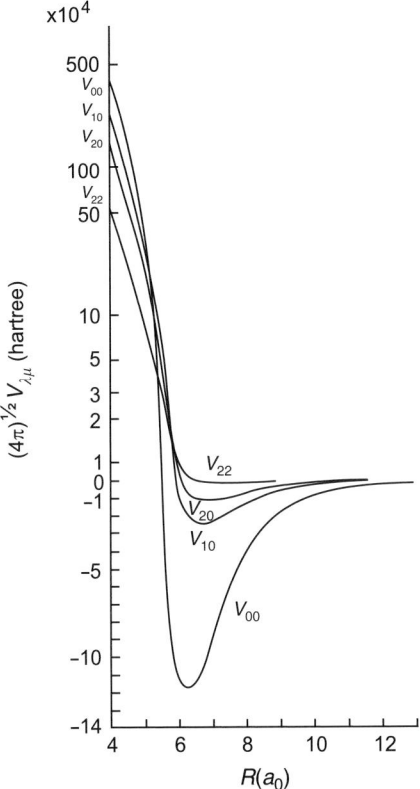

Figure 7.7 The computed values of the coefficients $(4\pi)^{\frac{1}{2}} V_{\lambda\mu}(R)$ of the expansion (7.41) of the interaction potential between OH and para-H_2.

to the invariance of V with respect to rigid rotations of the entire (bimolecular) system about the intermolecular axis. The relationship between the spherical harmonic that appears in (7.47) and the corresponding rotation matrix element is

$$Y_{\lambda_2,-\nu}(\theta_2,\phi_2) = \left(\frac{2\lambda_2+1}{4\pi}\right)^{\frac{1}{2}} D^{\lambda_2*}_{-\nu 0}(\phi_2,\theta_2,0) \tag{7.48}$$

Using the definition (4.14) of the rotation matrix elements, equation (7.47) becomes

$$V(R,\Theta,\Phi,\theta_2,\phi_2) = \sum_{\lambda_1\mu\lambda_2\nu} v_{\lambda_1\mu\lambda_2\nu}(R) \left(\frac{2\lambda_2+1}{4\pi}\right)^{\frac{1}{2}}$$
$$\times e^{i\mu\Phi} d^{\lambda_1}_{\mu\nu}(\Theta) e^{-i\nu\phi_2} d^{\lambda_2}_{-\nu 0}(\theta_2) \tag{7.49}$$

where we have made use of the fact that the $d^j_{m'm}(\beta)$ are real. Using *ab initio* calculations [177], the coefficients $v_{\lambda_1\mu\lambda_2\nu}(R)$ in the expansion (7.49) may be derived for the values of the indices in Table 7.2.

Table 7.2. *Indices of the coefficients* $v_{\lambda_1\mu\lambda_2\nu}(R)$ *of the expansion (7.49) of the* OH–H$_2$ *interaction potential, which may be derived from ab initio calculations [177]; '1' denotes the* OH *radical and '2' the* H$_2$ *molecule. Note that* $v_{\lambda_1\mu\lambda_2\nu}(R) \equiv v_{\lambda_1-\mu\lambda_2-\nu}(R)$.

λ_1	μ	λ_2	ν
0	0	0	0
1	0	0	0
2	0	0	0
0	0	2	0
1	0	2	0
2	0	2	0
2	0	2	2
2	2	0	0
2	2	1	0
2	2	2	0
2	2	2	2
2	2	2	-2

7.5.2 OH–H$_2$ collisions

The matrix elements of the interaction potential were evaluated by Offer [178], in the space-fixed (laboratory) coordinate system. The potential expansion coefficients in the space-fixed frame, $v_{\lambda_1\lambda_2\lambda\mu}(R)$, are related to the $v_{\lambda_1\mu\lambda_2\nu}(R)$ which appear in equation (7.49) by

$$v_{\lambda_1\lambda_2\lambda\mu}(R) = \left(\frac{4\pi}{2\lambda+1}\right)^{\frac{1}{2}} \sum_{\nu\geq 0} C^{\lambda_1\lambda_2\lambda}_{\nu-\nu 0}(1+\delta_{\nu 0})^{-1}$$
$$\times [v_{\lambda_1\mu\lambda_2\nu}(R) + (-1)^{\lambda_1+\lambda_2+\lambda}v_{\lambda_1\mu\lambda_2,-\nu}(R)] \qquad (7.50)$$

Using equation (7.50), the potential matrix elements become

$$\langle J_1\Omega\epsilon J_2 J_{12}lJM|V|J_1'\Omega'\epsilon'J_2'J_{12}'l'JM\rangle$$

$$= \sum_{\lambda_1\lambda_2\lambda}(-1)^{J_1'+J_2'-J_{12}-\Omega'-J}\left(\frac{2\lambda+1}{4\pi}\right)\frac{1-\epsilon\epsilon'(-1)^{J_1+J_1'+\lambda+\lambda_2}}{2}$$

$$\times [(2J_1+1)(2J_2+1)(2J_{12}+1)(2l+1)(2\lambda_2+1)(2l'+1)(2J_{12}'+1)(2J_2'+1)(2J_1'+1)]^{\frac{1}{2}}$$

$$\times \left[\begin{pmatrix}\lambda_1 & J_1 & J_1'\\ 0 & \Omega & -\Omega'\end{pmatrix}\delta_{\Omega\Omega'}v_{\lambda_1\lambda_2\lambda 0}(R) + \epsilon\begin{pmatrix}\lambda_1 & J_1 & J_1'\\ 2 & -\Omega & -\Omega'\end{pmatrix}(1-\delta_{\Omega\Omega'})v_{\lambda_1\lambda_2\lambda 2}(R)\right]$$

$$\times \begin{pmatrix}\lambda & l & l'\\ 0 & 0 & 0\end{pmatrix}\begin{pmatrix}\lambda_2 & J_2 & J_2'\\ 0 & 0 & 0\end{pmatrix}\begin{Bmatrix}l' & l & \lambda\\ J_{12} & J_{12}' & J\end{Bmatrix}\begin{Bmatrix}J_{12} & J_2 & J_1\\ J_{12}' & J_2' & J_1'\\ \lambda & \lambda_2 & \lambda_1\end{Bmatrix} \qquad (7.51)$$

7.5 Rotational excitation of OH by H_2

where '1' denotes the OH radical and '2' the H_2 molecule; Ω, Ω' denote the projection of J_1, J_1' on the symmetry (internuclear) axis of the radical; $\begin{pmatrix} \cdots \\ \cdots \end{pmatrix}$ is a Wigner 3j-, $\begin{Bmatrix} \cdots \\ \cdots \end{Bmatrix}$ a 6j-, and $\begin{Bmatrix} \cdots \\ \cdots \\ \cdots \end{Bmatrix}$ a 9j-coefficient. $\mathbf{J}_{12} = \mathbf{J}_1 + \mathbf{J}_2$ and $\mathbf{J} = \mathbf{J}_{12} + \mathbf{l}$, where $\mathbf{l}$ is the relative orbital angular momentum of the OH radical and the H_2 molecule and $\mathbf{J}$ is the total angular momentum of the system. It may be seen that the matrix elements (7.51) vanish unless

$$\epsilon\epsilon'(-1)^{J_1+J_1'+\lambda+\lambda_2} = -1 \tag{7.52}$$

where ϵ, ϵ' distinguish the components of a Λ-doublet (see Fig. 7.1). The form of the matrix elements (7.51) is adapted to the case a coupling scheme; the appropriate linear combinations of these matrix elements have to be taken when performing calculations in intermediate coupling [cf. equation (7.35)].

Introducing the sum of the initial and final values of the spin quantum number, $S = S' = 1/2$, into the identity (7.52), we obtain

$$\epsilon\epsilon'(-1)^{J_1+J_1'+\lambda+\lambda_2-S-S'} = +1 \tag{7.53}$$

Finally, noting that $\lambda + l + l'$ and $\lambda_2 + J_2 + J_2'$ must be even in order that the last two 3j-coefficients in equation (7.51) should not vanish identically, and that λ and λ_2 are integers, the condition for non-zero potential matrix elements becomes

$$\epsilon(-1)^{J_1-S+J_2+l} = \epsilon'(-1)^{J_1'-S'+J_2'+l'} \tag{7.54}$$

Recalling that the parity of the OH eigenstates is $(-1)^{J_1-S}\epsilon$, that the parity of the H_2 rotational functions is $(-1)^{J_2}$, and that the parity of the function describing the relative motion of the radical and the molecule is $(-1)^l$, we see that equation (7.54) is an expression of the conservation of the overall parity of the radical–molecule scattering system.

The only laboratory measurements to date of the cross-sections for the rotational excitation of OH by H_2 were performed, using 'normal' H_2 (a 3:1 mixture of ortho:para) and the 'crossed–beam' technique, at a collision energy $E = 83$ meV ($E/k_B = 963$ K) [179]. These measurements demonstrated that selective excitation of particular components of the Λ-doublets does, indeed, occur in OH–H_2 collisions, with the propensity being to anti-invert the populations of the Λ-doublets in the $^2\Pi_{\frac{3}{2}}$ ladder, and to invert the Λ-doublets with $J \leq 7/2$ in the $^2\Pi_{\frac{1}{2}}$ ladder. The laboratory measurements were in qualitative agreement with calculations [180, 181].

Offer and van Dishoeck [182] have made the most complete calculations to date of cross-sections and rate coefficients for the collisional excitation of OH by para- and ortho-H_2; their calculations were in intermediate coupling. The rate coefficients for transitions induced by ortho-H_2 ($J = 1$) are larger, owing to the non-vanishing interactions with the dipole moment of the OH radical [which are absent for collisions with para-H_2 ($J = 0$)]. Assuming that the ratio of ortho:para H_2 was thermalized at the kinetic temperature of the warm ($T \approx 100$ K) gas, Offer and van Dishoeck found that the inclusion of collisions with ortho-H_2 had the effect of *reducing* their selectivity with respect to populating the components of the Λ-doublets. Thus, the propensity of collisions with H_2 to anti-invert the populations of the Λ-doublets in the $^2\Pi_{\frac{3}{2}}$ ladder of OH, and to invert the Λ-doublets with $J \leq 7/2$ in the $^2\Pi_{\frac{1}{2}}$ ladder, is reduced when allowance is made for the presence of ortho-H_2 in the medium.

Offer and van Dishoeck [182] also computed the intensities of the rotational transitions of OH, under conditions of their excitation in thermal collisions with H_2. These transitions were subsequently observed by the *Infrared Space Observatory* (ISO).

While collisional excitation by H_2 molecules cannot give rise directly to masing in the Λ-doublet transitions within the $^2\Pi_{\frac{3}{2}}$ ladder of OH, the process of *radiative* excitation by infrared photons followed by *de*-excitation in collisions with H_2 molecules can lead to population inversion within these Λ-doublets. The density of the gas has to be high, in excess of about 10^6 cm^{-3}, for collisional de-excitation to become competitive with radiative decay, owing to the large values of the spontaneous radiative transition probabilities [182]. Furthermore, the temperature of the grains, which emit the infrared radiation, has to exceed the kinetic temperature of the gas. This mechanism may be significant in the case of OH masers observed in association with H II regions.

8
Charge transfer processes

8.1 Introduction

The process of charge transfer (charge exchange) plays an important role in the ionization balance of the interstellar gas and in the formation of molecules in the gas phase. For example, the forwards and reverse reactions

$$H^+ + O \rightleftharpoons H + O^+ \quad (8.1)$$

are the key to the O^0–O^+ ionization equilibrium in planetary nebulae and supernovae remnants. This reaction is also the cornerstone of the ion–molecule chemistry leading to the formation of oxygen-bearing molecules in diffuse clouds (see sub-section entitled 'Diffuse clouds' in Section 1.2.2). Other examples are charge transfer reactions of the type

$$H_3O^+ + Fe \rightarrow Fe^+ + H_2O + H \quad (8.2)$$

which lead to the ionization of 'metals' such as Mg, Si and Fe. Such processes are particularly important in dark clouds, from whose interiors the Galactic background ultraviolet radiation field is excluded by dust absorption. Electrons tend to be transferred from atoms with low ionization potentials (Mg: 7.65 eV; Si: 8.15 eV; Fe: 7.87 eV), whose ions neutralize only slowly, through radiative recombination with electrons.

Charge transfer involving an ion and an atom or an ion and a molecule may be discussed, especially at low energies, in terms of the quasi-molecule that is temporarily formed during the collision. To a first approximation, the target and projectile may be considered to move relative to each other along an adiabatic potential energy curve (for a collision between an ion and an atom) or surface (for a collision between an ion and a molecule). However, if charge transfer is to take place, a transition to a *different* potential energy curve (or surface) must occur. At large separations of the interacting particles, these potential energy curves differ by an energy that is equal to the difference in the ionization potentials of the particles involved. For example, in the reaction (8.1), the relevant potential energy curves correlate at large internuclear separation with $H^+ + O$, on the one hand, and with $H + O^+$, on the other hand. The difference in the ionization potentials of O and H is $I(O) - I(H) = 13.618 - 13.598 = 0.020$ eV ($\equiv 227$ K), for O in its $2p^4\ ^3P_2$ ground state. Thus, reaction (8.1) is endothermic by 227 K in the forwards direction. As the excited fine structure level O $2p^4\ ^3P_1$ lies 228 K above the ground $2p^4\ ^3P_2$ level, charge transfer from the first excited level to H^+ is almost resonant. However, only at high densities $[n(H_2) > 10^5\ \text{cm}^{-3}]$ is the fractional population of the excited level sufficiently large for this process to be significant.

140   *Charge transfer processes*

If one of the interacting particles is left in an excited state subsequent to charge transfer, the potential energy curve that correlates with that state is implicated. Charge transfer into excited states of the product atom is particularly important in reactions of hydrogen with multiply charged ions

$$X^{m+} + H \rightarrow [X^{(m-1)+}]^* + H^+ \tag{8.3}$$

in which $m \geq 2$. Thus, charge transfer at low energies is a molecular collision process involving more than one potential energy curve (or surface). In this respect, charge transfer is similar to the types of collision process involving open-shell atoms and molecules that were considered in Chapters 6 and 7.

We shall discuss first the Landau–Zener model of charge transfer. This model is helpful in visualizing the process and in illustrating the important concepts of 'adiabatic' and 'diabatic' interactions. The model of 'orbiting' in low-energy collisions between an ion and a neutral will also be found to be useful; this phenomenon was introduced in Chapter 6, in connection with fine structure excitation processes. Finally, we consider charge transfer at low energies from the quantum mechanical standpoint, combining a molecular orbital representation of the electronic wave function with a wave mechanical treatment of the relative motion of the projectile and the target.

8.2   The Landau–Zener model

In order to illustrate the principles involved, we shall consider the reaction

$$X^{m+} + H \rightarrow X^{(m-1)+} + H^+ + \Delta E \tag{8.4}$$

where $\Delta E = I(X^{(m-1)+}) - I(H)$ is the difference between the ionization potentials of the ion $X^{(m-1)+}$ and atomic hydrogen. In the initial channel, $X^{m+} + H$, the dominant long-range interaction between the target and the projectile is the polarization potential

$$V_{\text{pol}} = -\frac{\alpha m^2}{2R^4} \tag{8.5}$$

where $\alpha = 4.5$ is the polarizability of atomic hydrogen, in atomic units, and R is the separation of the target and projectile. For the purposes of the discussion, we shall consider the case $m \geq 2$, where the dominant long-range force in the final channel, $X^{(m-1)+} + H^+$, is the coulomb repulsion

$$V_{\text{coul}} = \frac{m-1}{R} \tag{8.6}$$

in atomic units. These two potential energy curves are sketched in Fig. 8.1. In the diagram, the curve representing the coulomb interaction (8.6) has been shifted downwards by an amount ΔE, the energetic separation of the initial and final channels, so that the relative asymptotic energies (as $R \rightarrow \infty$) are given correctly.

Let us suppose that the total electronic spin, S, and the magnitude of the projection, Λ, of the total electronic orbital angular momentum on the internuclear axis are the same in both the initial and the final channels. In this case, the molecular states and the associated potential energy curves are said to have the same symmetry, $^{2S+1}\Lambda$, where $\Lambda = 0, 1, 2, \ldots$ are denoted

8.2 The Landau–Zener model

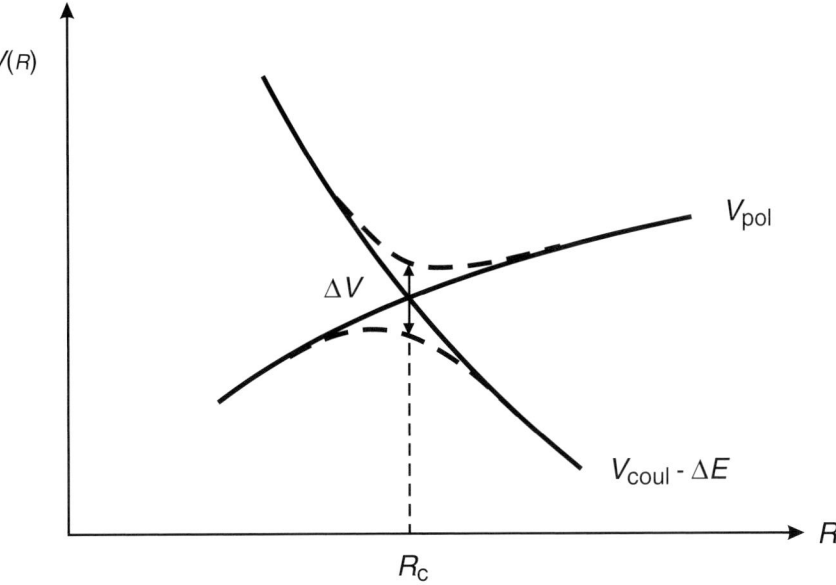

Figure 8.1 The polarization potential between X^{m+} and H and the coulomb potential between $X^{(m-1)+}$ and H^+, showing the avoided crossing at $R = R_c$, where R is the internuclear distance. The continuous curves are the *diabatic* potentials, and the broken curves are the *adiabatic* potentials.

$\Sigma, \Pi, \Delta, \ldots$. Under these circumstances, and for reasons to be given below, the *adiabatic* potential energy curves do not cross; there is instead an 'avoided crossing', as indicated by the dashed lines in Fig. 8.1. The avoided crossing is at $R = R_c$, where R_c is the solution of the equation

$$V_{\text{coul}}(R_c) - V_{\text{pol}}(R_c) = \Delta E \qquad (8.7)$$

At sufficiently long range, $V_{\text{coul}} \gg V_{\text{pol}}$, and then

$$\frac{m-1}{R_c} \approx \Delta E$$

or

$$R_c \approx \frac{m-1}{\Delta E} \qquad (8.8)$$

in atomic units. Thus, when Si^{2+} undergoes charge transfer with atomic hydrogen, yielding Si^+, whose ionization potential exceeds that of hydrogen by 2.74 eV, the avoided crossing is located at $R \approx 10\, a_0$, where a_0 is the atomic unit of length (the Bohr radius).

Let us denote by $\Delta V(R)$ the energy separation of the adiabatic potential energy curves (see Fig. 8.1). In order for charge transfer to occur, a transition ('jump') must take place from one potential energy curve to another. Physical intuition suggests that such a transition is most probable in the vicinity of the avoided crossing at $R = R_c$. An expression for the probability of the transition may be derived within the framework of the Landau–Zener model.

Charge transfer processes

The derivation begins with the consideration of the relevant time scales. The characteristic time, τ, of the internal motion responsible for the transition is given by

$$\tau \approx 1/\Delta V(R) \tag{8.9}$$

in atomic units. The strength of the coupling between the *diabatic* (continuous) curves in Fig. 8.1 determines the magnitude of the separation, ΔV, of the adiabatic (broken) curves at the avoided crossing. A direct comparison may be drawn with the analysis of the effect of a perturbation on the energy levels of atoms. Levels of the same symmetry 'repel' each other, with the magnitude of the repulsion being, according to second-order perturbation theory, inversely proportional to the magnitude of their energetic separation, calculated to first order [183].

A characteristic time of transit through the avoided crossing may be defined as the time required for the target and projectile to move a distance ΔR from $R = R_c$, such that the separation $\Delta V(R_c \pm \Delta R)$ of the adiabatic curves is twice its minimum value, $\Delta V(R_c)$. A Taylor series expansion about $R = R_c$ then yields

$$\frac{\Delta V(R_c)}{\Delta R} \approx |V'_{coul}(R_c) - V'_{pol}(R_c)| \tag{8.10}$$

where the primes indicate differentiation with respect to R. The collision time, as defined above, is given by

$$T \approx \Delta R / v_l(R_c)$$
$$\approx \frac{\Delta V(R_c)}{v_l(R_c)|V'_{coul}(R_c) - V'_{pol}(R_c)|} \tag{8.11}$$

in which $v_l(R_c)$ is the relative speed of the target and projectile at the avoided crossing for a given relative orbital angular momentum, l. Comparing these two characteristic times, we obtain

$$\frac{T}{\tau} \approx \frac{(\Delta V)^2}{v_l |V'_{coul} - V'_{pol}|} \tag{8.12}$$

where all quantities on the right-hand side are to be evaluated at $R = R_c$. In a slow collision, $T/\tau \gg 1$ and the particles tend to follow the adiabatic potential energy curves in Fig. 8.1. When the collision is fast, $T/\tau \ll 1$ and the particles follow the diabatic potential curves. In both of these limits, the probability of charge transfer is small: if the interacting particles traverse $R = R_c$, they do so twice, on the way in and on the way out, and, at the end of the collision, they tend to be on the same potential energy curve and in the same state as at the start.

The Landau–Zener model yields an expression for the probability, P_l, that a jump is made between the adiabatic potential energy curves in a *single* transit through the avoided crossing. The expression is

$$P_l = e^{-\omega} \tag{8.13}$$

where

$$\omega = \frac{\pi}{2} \frac{T}{\tau} \tag{8.14}$$

8.3 The 'orbiting' model

The total probability of charge transfer is

$$\mathcal{P}_l = P_l(1 - P_l) + (1 - P_l)P_l \tag{8.15}$$

where the first term on the right-hand side is the product of the probability that a transition to the other potential energy curve occurs on the inward trajectory and that the particles stay on the same potential energy curve on the outward trajectory; the second term on the right-hand side is the contribution from the analogous process in which the transition to the other potential curve occurs on the way out. The cross-section, σ, for charge transfer may be obtained from the integral of the probability over the impact parameter, b,

$$\sigma = 2\pi \int_0^\infty \mathcal{P}(b) b \, \mathrm{d}b \tag{8.16}$$

to which there is an equivalent summation over l

$$\sigma = \frac{\pi}{k_i^2} \sum_{l=0}^\infty (2l+1) \mathcal{P}_l \tag{8.17}$$

where k_i is the wave number in the incident channel, i. In practice, the infinite summation in equation (8.17) terminates at $l = L$, the classical turning point which is such that $v_L(R_c) = 0$. For $l > L$, $R > R_c$ and the avoided crossing is not reached.

8.3 The 'orbiting' model

Bates [184] showed that $E_i(R_c) \gg \Delta V(R_c)$ is a necessary condition for the Landau–Zener approximation to be valid, where $E_i(R_c)$ is the relative kinetic energy at the avoided crossing in the incident channel. In other words, the Landau–Zener approximation is adapted to high collision energies; at low energies, the Landau–Zener formula tends to underestimate seriously the charge transfer cross-section.

The failure of the Landau–Zener approximation at low energies is partly attributable to the neglect of classical trajectory effects. The classical concept of 'orbiting' has already been introduced in Chapter 6, in the context of the collisional excitation of fine structure transitions in atoms and ions. The same model will now be applied to the charge transfer process (8.4), at low collision energies.

The classical equation of energy conservation states that

$$E_i = \frac{1}{2}\mu v_l^2(R) + \frac{l^2}{2\mu R^2} - \frac{\alpha m^2}{2R^4} \tag{8.18}$$

where E_i is the total energy, μ is the reduced mass, and l is the relative angular momentum. The total energy, E_i is equal to the incident kinetic energy at infinite separation of the target and projectile. The terms on the right-hand side of equation (8.18) are: the kinetic energy associated with the radial component of the relative velocity; the kinetic energy associated with the angular component of the relative velocity; and the potential energy. Using the relation $l = (2\mu E_i)^{\frac{1}{2}} b$ between the angular momentum and the classical impact parameter, b, equation (8.18) becomes

$$E_i = \frac{1}{2}\mu v_l^2(R) + \frac{E_i b^2}{R^2} - \frac{\alpha m^2}{2R^4} \tag{8.19}$$

At the point of closest approach, $R = R_0$, $v_l(R_0) = 0$ and hence

$$E_i = \frac{E_i b^2}{R_0^2} - \frac{\alpha m^2}{2R_0^4} \tag{8.20}$$

Orbiting of the target and projectile occurs when the centrifugal and polarization forces balance, that is,

$$\frac{2E_i b^2}{R_0^3} = \frac{2\alpha m^2}{R_0^5} \tag{8.21}$$

Eliminating R_0 between equations (8.20) and (8.21), we obtain the critical value of the impact parameter at which orbiting takes place:

$$b = \left(\frac{2\alpha m^2}{E_i}\right)^{\frac{1}{4}} \tag{8.22}$$

According to the classical picture, the target and projectile remain close together for a relatively long period of time during the orbiting motion. It is to be anticipated that the probability of charge transfer is thereby enhanced. If $P = \frac{1}{2}$ for such a collision, then $\mathcal{P} = \frac{1}{2}$, its maximum value, and the charge transfer cross-section is given by

$$\sigma \approx \frac{1}{2}\pi \left(\frac{2\alpha m^2}{E_i}\right)^{\frac{1}{2}} \tag{8.23}$$

In Table 8.1, the values of the cross-section derived from equation (8.23) are compared with the results of quantum mechanical calculations, for the exothermic reaction

$$\mathrm{Si}^{2+} + \mathrm{H} \rightarrow \mathrm{Si}^+ + \mathrm{H}^+ + 32\,000\,\mathrm{K} \tag{8.24}$$

in which $m = 2$, and the exothermicity of the reaction, 2.74 eV, has been expressed in K. The comparison in Table 8.1 confirms the validity of the orbiting model at low collision energies.

The cross-section (8.23) is inversely proportional to the incident velocity of the projectile. It follows that the rate coefficient, $\langle \sigma v \rangle$, is a constant, independent of the kinetic temperature, T. The rate coefficient is defined as

$$\langle \sigma v \rangle = \left(\frac{8k_\mathrm{B}T}{\pi\mu}\right)^{\frac{1}{2}} \int_0^\infty x\sigma(x) e^{-x}\, dx \tag{8.25}$$

where k_B is Boltzmann's constant and $x = E_i/(k_\mathrm{B}T)$. Using equation (8.23), the integral in (8.25) may be evaluated, yielding

$$\langle \sigma v \rangle = m\pi (\alpha/\mu)^{\frac{1}{2}} \tag{8.26}$$

This equation is an expression of the 'Langevin' rate coefficient for exothermic ion-neutral reactions, of which charge transfer is one example.

8.4 The quantum mechanical model

Table 8.1. *Cross-sections, σ, computed for the charge transfer reaction (8.24) as functions of the incident collision energy, E_i: (1) quantum mechanical calculations [185]; (2) prediction of the orbiting approximation, equation (8.23).*

E_i/k_B (K)	$\sigma^{(1)}(a_0^2)$	$\sigma^{(2)}(a_0^2)$
1	6980	5296
3	3730	3058
10	1570	1675
30	826	967
100	472	530
300	259	306
1000	197	167

8.4 The quantum mechanical model

At the low kinetic energies which occur in many parts of the interstellar medium, only a fully quantum mechanical calculation may be expected to yield a reliable solution to the problem of charge transfer. Accordingly, we shall now consider the use of quantum mechanics to solve such problems.

8.4.1 Formulation

By analogy with the discussion of the Born–Oppenheimer approximation in Chapter 4, let us consider charge transfer involving a one-electron atom, A, and a fully-stripped ion, B. The atom A may be considered to be moving with reduced mass $\mu = m_A m_B/(m_A + m_B)$ relative to a fixed centre of force, B. The internuclear coordinates will be denoted by $\mathbf{R}$ and the position vector of the electron with respect to the centre of mass of A and B by $\mathbf{x}$.

It was shown in Chapter 4 that a solution, ψ, of the Schrödinger equation

$$H\psi = E\psi \qquad (8.27)$$

may be sought in the form

$$\psi(\mathbf{x}, \mathbf{R}) = \sum_i F_i(\mathbf{R})\phi_i(\mathbf{x}, \mathbf{R}) \qquad (8.28)$$

where the functions ϕ_i are themselves solutions of an eigenvalue equation

$$\left[H(\mathbf{x}, \mathbf{R}) + \frac{\nabla_R^2}{2\mu}\right]\phi_i(\mathbf{x}, \mathbf{R}) = E_i(\mathbf{R})\phi_i(\mathbf{x}, \mathbf{R}) \qquad (8.29)$$

When the relative velocity of the nuclei is large, it is desirable to incorporate 'electron translation factors' in the expansion (8.28) of the wave function. These factors allow explicitly for the motion of the electron of atom A relative to the centre of mass of A and B. Following the original work of Bates and McCarroll [186], plane-wave translation factors are often employed,

$$\exp\left[i(p\mathbf{v}\cdot\mathbf{x} - \frac{1}{2}p^2v^2t)\right]$$

where $i = (-1)^{\frac{1}{2}}$, $p = m_B/(m_A + m_B)$ is the fractional distance of A from the centre of mass of A and B, and $\mathbf{v}$ is the relative velocity of A and B. The inclusion of translation factors ensures that the solutions of equation (8.27) have the correct form as $R \to \infty$. We shall consider the low-velocity regime, in which translation factors are dispensable.

Since $-\nabla_R^2/(2\mu)$ represents the relative nuclear kinetic energy, $[H(\mathbf{x}, \mathbf{R}) + \nabla_R^2/(2\mu)]$ contains all contributions to the total hamiltonian, H, except that corresponding to the relative nuclear motion. As shown in Section 4.2, Schrödinger's equation reduces to

$$\left[-\frac{\nabla_R^2}{2\mu} + E_j(\mathbf{R}) - E\right] F_j(\mathbf{R})$$
$$= \sum_i \left[\frac{\langle\phi_j|\nabla_\mathbf{R}|\phi_i\rangle\cdot\nabla_\mathbf{R} F_i(\mathbf{R})}{\mu} + \frac{\langle\phi_j|\nabla_R^2|\phi_i\rangle F_i(\mathbf{R})}{2\mu}\right] \quad (8.30)$$

where the terms on the right-hand side of equation (8.30) arise from the coupling of the electronic and nuclear motions. The Born–Oppenheimer approximation consists of neglecting these couplings. In this case, an adiabatic potential energy curve is followed during the collision, and charge transfer cannot occur. Like the excitation of fine structure transitions, considered in Chapter 6, charge transfer arises from the coupling of the electronic to the nuclear motion and is an essentially 'non-adiabatic' process.

At low collision energies, it is appropriate to consider the body-fixed (BF) coordinate frame, in which the Z-axis is taken to coincide with the internuclear vector, $\mathbf{R}$. The orientation of the BF Z-axis relative to the space-fixed (SF) frame is (Θ, Φ), where $\mathbf{R} = (R, \Theta, \Phi)$ are spherical polar coordinates relative to the SF frame (see Fig. 8.2). An alternative expression for the wave function, ψ, is

$$\psi(\mathbf{x}, \mathbf{R}) = \left(\frac{2J+1}{4\pi}\right)^{\frac{1}{2}} \sum_{\alpha, \Lambda} \frac{G(\alpha\Lambda|R)}{R} D_{M\Lambda}^{J*}(\Phi, \Theta, 0) \chi(\alpha, \Lambda|\mathbf{x}', R) \quad (8.31)$$

in which $D_{M\Lambda}^{J*}(\Phi, \Theta, 0)$ denotes the complex conjugate of an element of the rotation matrix [48] [a rotation through the Euler angles $(\Phi, \Theta, 0)$ takes the SF into the BF frame]; $\chi(\alpha, \Lambda|\mathbf{x}', R)$ is the electronic eigenfunction, the electronic coordinates $\mathbf{x}'$ being expressed relative to the BF frame. The electronic state is characterized by the magnitude of the projection, Λ, of the electronic angular momentum, $\mathbf{L}$, on the BF Z-axis, and the other quantum numbers required to specify the state completely are denoted by α. The total angular momentum, $\mathbf{J}$, of the quasi-molecule AB is the vector sum of $\mathbf{L}$ and the angular momentum associated with the relative motion of the nuclei; M is the projection of $\mathbf{J}$ on the SF z-axis.

The form taken by the scattering equations when the expansion (8.31) is substituted into Schrödinger's equation (8.27) has been considered in detail by Gaussorgues et al. [187]. Schrödinger's equation reduces to a set of coupled, ordinary differential equations for the

8.4 The quantum mechanical model

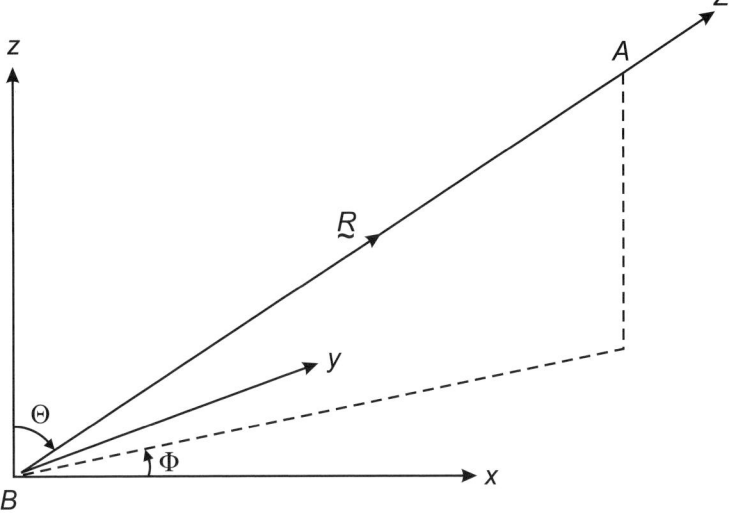

Figure 8.2 The spherical polar angles (Θ, Φ) which determine the orientation of the internuclear vector, **R** (which is taken to be the BF Z-axis), relative to the (x, y, z) SF frame.

expansion coefficients, $G(\alpha\Lambda|R)$:

$$\left[\frac{d^2}{dR^2} - \frac{J(J+1) - \Lambda^2}{R^2} + 2\mu[E - E(\alpha\Lambda|R)]\right] G(\alpha\Lambda|R)$$

$$= -\sum_{\alpha'} 2\left\langle\alpha\Lambda\left|\frac{\partial}{\partial R}\right|\alpha'\Lambda\right\rangle \frac{d}{dR} G(\alpha'\Lambda|R) + \left\langle\alpha\Lambda\left|\frac{\partial^2}{\partial R^2}\right|\alpha'\Lambda\right\rangle G(\alpha'\Lambda|R)$$

$$+ \frac{2}{R^2}[(J \mp \Lambda)(J \pm \Lambda + 1)]^{\frac{1}{2}} \langle\alpha\Lambda|L_{y'}|\alpha', \Lambda \pm 1\rangle G(\alpha', \Lambda \pm 1|R) \quad (8.32)$$

In equation (8.32), $|\alpha\Lambda\rangle \equiv \chi(\alpha\Lambda|\mathbf{x}', R)$ and $E(\alpha\Lambda|R)$ is the corresponding adiabatic potential energy curve.

The first two 'radial coupling' terms on the right-hand side of equation (8.32) are responsible for charge transfer in the case $\Lambda = \Lambda'$. Of these two terms, the first is usually more important than the second. The last term on the right-hand side arises from 'rotational coupling', i.e. from the rotation of the BF relative to the SF coordinate system; this coupling induces transitions between potential curves of angular symmetry Λ and $\Lambda' = |\Lambda \pm 1|$. $L_{y'}$ is the y-component of the relative electronic angular momentum operator, in the BF frame [187]. The more important case, in practice, corresponds to the initial and final potential energy curves having the same symmetry ($\Lambda = \Lambda'$); this gives rise to an avoided crossing, as illustrated in Fig. 8.1. In other words, rotational coupling is generally less important than radial coupling, and only the latter will be considered below.

The leading term on the right-hand side of equation (8.32) is

$$2\left\langle\alpha\Lambda\left|\frac{\partial}{\partial R}\right|\alpha'\Lambda\right\rangle \frac{d}{dR} G(\alpha'\Lambda|R)$$

This matrix element involves the electronic wave functions, $\langle\alpha\Lambda|$ and $|\alpha'\Lambda\rangle$, which vary rapidly with R in the vicinity of an avoided crossing. Indeed, it is the *dynamic* (radial) coupling between the molecular states of the same symmetry which causes the adiabatic (broken) curves to 'avoid' each other in the vicinity of $R = R_c$ (cf. Fig. 8.1). An alternative viewpoint is to consider the projectile and target to move relative to each other on the diabatic (continuous) curves in Fig. 8.1, in which case charge transfer occurs through *potential* coupling, as will be seen below.

The expansion coefficents in the *adiabatic* basis, $G(\alpha\Lambda|R)$, may be related to those in a *diabatic* basis, $\mathcal{G}(\beta\Lambda|R)$, through the transformation

$$G(\alpha\Lambda|R) = \sum_\beta U(\alpha\Lambda, \beta\Lambda|R)\mathcal{G}(\beta\Lambda|R) \tag{8.33}$$

where $U(\alpha\Lambda, \beta\Lambda|R)$ is an element of a unitary transformation matrix. In the case of *two* interacting molecular states, which we are considering, the radial coupling matrix elements, $W(\alpha, \alpha') = \langle\alpha\Lambda|\frac{\partial}{\partial R}|\alpha'\Lambda\rangle$, satisfy the relationships [188]

$$W(1,1) = W(2,2) = 0$$
$$W(1,2) = -W(2,1) \tag{8.34}$$

The corresponding 2×2 transformation matrix, $\mathbf{U}(R)$, then takes the form

$$\mathbf{U}(R) = \begin{pmatrix} \cos\omega(R) & \sin\omega(R) \\ -\sin\omega(R) & \cos\omega(R) \end{pmatrix} \tag{8.35}$$

where

$$\omega(R) = \int_R^\infty \left\langle 1\Lambda \left| \frac{\partial}{\partial R'} \right| 2\Lambda \right\rangle dR' \tag{8.36}$$

and $\mathbf{U}(R)$ satisfies the differential equation

$$\frac{d}{dR}\mathbf{U}(R) + \mathbf{W}(R)\mathbf{U}(R) = 0 \tag{8.37}$$

in which $\mathbf{W}(R)$ is the radial coupling matrix whose elements are given by equation (8.34). Using (8.32), (8.33) and (8.37), it may be shown that the expansion coefficients in the diabatic basis, $\mathcal{G}(\beta\Lambda|R)$, satisfy the coupled differential equations

$$\left[\frac{d^2}{dR^2} - \frac{J(J+1) - \Lambda^2}{R^2} + k^2\right]\mathcal{G}(\beta\Lambda|R)$$
$$= 2\mu \sum_{\beta'} V(\beta, \beta'|R)\mathcal{G}(\beta'\Lambda|R) \tag{8.38}$$

where $k^2 = 2\mu E$, and the potential matrix, $\mathbf{V}(R)$, takes the form

$$\mathbf{V}(R) = \begin{pmatrix} E_1\cos^2\omega + E_2\sin^2\omega & (E_1 - E_2)\sin\omega\cos\omega \\ (E_1 - E_2)\cos\omega\sin\omega & E_1\sin^2\omega + E_2\cos^2\omega \end{pmatrix} \tag{8.39}$$

where $E_\alpha \equiv E(\alpha\Lambda|R)$ and $\omega \equiv \omega(R)$.

8.4 The quantum mechanical model

Equation (8.36) shows that, as $R \to \infty$, $\omega(R) \to 0$, and hence, from equation (8.35), $\mathbf{U}(R) \to \mathbf{1}$, the unit matrix. It follows that, at large internuclear separations, the adiabatic and diabatic states and their associated potential energy curves are identical. This identity would remain true for all internuclear distances if the radial coupling was vanishingly small for all R. In general, the off-diagonal ($\beta \neq \beta'$) elements of the potential matrix, $\mathbf{V}(R)$, are finite and induce transitions between the diabatic states, leading to charge transfer.

The equations (8.38) have the standard form encountered, for example, when discussing the rotational excitation of molecules in collisions [Chapter 4, equations (4.57) and (4.65)] and may be solved using standard numerical techniques. From the solution in the asymptotic region (large R) may be extracted the scattering matrix, $\mathbf{S}$, and the charge transfer cross-sections. Rate coefficients, $\langle \sigma v \rangle$, as functions of the kinetic temperature of the gas, T, are derived from the cross-sections, σ, as functions of the barycentric collision energy, E, by integrating over a Maxwellian velocity distribution.

8.4.2 Calculations

The first and major difficulty encountered in charge transfer calculations is the determination of the molecular electronic states in the BF frame, $\chi(\alpha\Lambda|\mathbf{x}', R)$, and the associated adiabatic potential energy curves, $E(\alpha\Lambda|R)$. From these data, the potential matrix elements may be evaluated (in the case of two interacting molecular states) using (8.39).

As may be seen from equation (8.39), the diagonal elements of $\mathbf{V}(R)$, V_{11} and V_{22}, oscillate between $V_{11} = E(1\Lambda|R)$, $V_{22} = E(2\Lambda|R)$ and $V_{11} = E(2\Lambda|R)$, $V_{22} = E(1\Lambda|R)$ as $\omega(R)$ varies. Thus, when the coupling is strong and $\omega(R)$ varies rapidly with R, the diabatic curves will cross many times [189]. The off-diagonal elements of $\mathbf{V}(R)$, which are responsible for charge transfer, depend on the difference between the adiabatic energies, $E(1\Lambda|R) - E(2\Lambda|R)$. This energy difference can be difficult to evaluate accurately, as it is the difference between two relatively large numbers.

Table 8.2 summarizes the results of astrophysically relevant charge transfer calculations. Most of the calculations relate to charge transfer between ions and atomic hydrogen, in which the ion captures an electron; this is a *recombination* process for the ion concerned, whose rate is to be compared with the rates of radiative or dielectronic recombination (see Chapter 9) of the ion with free electrons. The rates (s^{-1}) of these processes are given by the products of the appropriate rate coefficients (cm^3 s^{-1}) with the H (or He) number density (cm^{-3}), or the free electron density (cm^{-3}), in the case of recombination with electrons. Thus, the relative importance of charge transfer recombination depends on the fractional ionization of the medium in question; but even in photoionized regions of the interstellar gas (planetary nebulae, H II regions, supernovae remnants), where the atomic hydrogen and helium densities are much lower than the electron density, charge transfer can be a significant and sometimes the dominant recombination process. The reason is that the charge transfer recombination reactions that we are discussing are *non*-radiative, whereas recombination with free electrons is a radiative process. The strength of the coupling between matter and the electromagnetic radiation field is proportional to the fine structure constant, $\alpha = e^2/(\hbar c) \approx 1/137$ [196], and this factor determines the relative orders of magnitude of comparable radiative and non-radiative processes (with the former being slower than the latter).

Table 8.2. *Calculated values of the rate coefficients, in units of 10^{-9} cm^3 s^{-1}, for charge transfer reactions with H^0 (or He^0, where indicated). Numbers in parentheses are powers of 10.*

Ion	T = 10	30	50	100	300	1000	3000 K	Reference
C^{3+}		1.5		1.6	1.6	1.6	1.6	[191]
C^{4+}	3.39	3.12		2.71	2.38	2.25	2.19	[192]
N^+				0.60(−3)		1.21(−3)		[193]
N^{3+}	0.35	0.30		0.25	0.25	0.43	1.12	[192]
N^{3+}		0.17		0.18	0.20	0.44	1.2	[191]
O^+	0.34		0.37	0.41		0.75		[194]
Si^{2+}	1.98		1.75	1.72		2.50		[195]

Ion	T = 5000	10 000	50 000 K	Reference
C^{2+}	1.00(−3)	1.00(−3)	1.49(−2)	[190]
C^{3+}	3.09	3.58	5.46	[190]
C^{3+}		1.6		[191]
C^{4+}		2.13	3.22	[192]
N^+	1.23(−3)	1.04(−3)	0.72(−3)	[193]
N^{2+}	0.78	0.86	1.11	[190]
N^{3+}	1.54	2.93	9.47	[190]
N^{3+}	1.82	3.41	11.23	[192]
N^{3+}		3.5		[191]
O^{2+}	0.60	0.77	1.62	[190]
$O^{2+} + He^0$	0.10	0.20	0.89	[190]
O^{3+}	6.34	8.63	17.6	[190]
Ne^{3+}	4.00	5.68	13.0	[190]
Si^{2+}	4.34	5.28	7.70	[195]

Under certain physical conditions, charge transfer may also act as an *ionization* process for an ion, as in the reverse of reaction (8.24):

$$Si^+ + H^+ \rightarrow Si^{2+} + H - 32\,000\,K \qquad (8.40)$$

Because of the large endothermicity involved, reaction (8.40) is significant only at high temperatures, such as the region of transition from the solar chromosphere to the corona [197] and in planetary nebulae. On the other hand, the reaction

$$Mg + H^+ \rightarrow Mg^+ + H + 18\,000\,K \qquad (8.41)$$

in which the Mg atom reacts in its $3s^2$ ^1S ground state, whereas the Mg^+ ion is produced in its 3p $^2P^o$ first excited state, is exothermic and proceeds even more rapidly than (8.40) in planetary nebulae [198], where $T \approx 10\,000$ K. The rate coefficients for reactions (8.40) and (8.41) are given in Table 8.3.

The cross-section for reaction (8.41) is strongly affected by resonances in the range of collision energies that is relevant at the kinetic temperatures which prevail in planetary

8.5 Selective population of excited states

Table 8.3. *Rate coefficients, in units of 10^{-9} cm^3 s^{-1}, for charge transfer reactions with H$^+$ which result in ionization of the specified reactant. Data from [198] for Mg and from [195] for Si$^+$.*

Reactant	$T = 5000$	10^4	20 000	50 000	10^5 K
Mg	0.033	0.188	0.589		
Si$^+$	0.003	0.073	0.426	1.360	2.207

nebulae. The resonances arise from coupling to a third molecular state of the MgH$^+$ system. Reaction (8.41) will be considered further in Section 8.5.

8.5 Selective population of excited states

From photoionized or collisionally ionized regions of the interstellar medium are observed spectra that include emission lines from a variety of atoms and ions. The excited states may be populated through electron collisional excitation, radiative or dielectronic recombination (see Chapter 9), but also by charge transfer, which leads to the formation of *excited* states of the product atom or ion. The relative importance of these processes depends on the fractional ionization of the medium. It follows that this important parameter – the degree of ionization of the gas – might be determined through observations of emission lines to whose intensities charge transfer makes a significant or a dominant contribution. An example of just such a process was considered in the previous section, namely

$$\text{Mg}(3s^2~^1S) + H^+ \rightarrow \text{Mg}^+(3p~^2P^o) + H(1s~^2S) + 18~000~K$$

in which the Mg$^+$ ion is produced in its 3p ^{2}P^o first excited state.

Another example of a selective excitation process is

$$O(2p^4~^3P) + H(1s~^2S) \rightarrow O(2p^4~^1D) + H(1s~^2S) - 22~800~K \quad (8.42)$$

which gives rise to the [O I] ^{1}D $\rightarrow ^3$P transition at 630 nm. This reaction proceeds through electron *exchange*, rather than single electron (charge) transfer, but involves similar mechanisms. The ^{1}D $\rightarrow ^3$P transition can be excited also by collisions with free electrons and, in regions where the degree of ionization of the gas is high, this latter process dominates. On the other hand, where the gas is only partially ionized, as may be the case in the shock-heated gas of Herbig–Haro (HH) objects, which are associated with regions of star formation, the charge transfer reaction (8.42) becomes significant; this reaction is endoergic by 1.97 eV (approximately 22 800 K). Federman and Shipsey [199] used the Landau–Zener approximation to calculate the rate coefficient for the reverse of reaction (8.42) for temperatures in the range $10^3 \leq T \leq 10^4$ K; their results are given in Table 8.4. The rate coefficients for the forward and reverse reactions, k_f and k_r, are related through the detailed balance relation

$$\omega_f k_f = \omega_r k_r \exp\left[-\frac{22~800}{T}\right] \quad (8.43)$$

Charge transfer processes

Table 8.4. *Rate coefficients, in units of 10^{-12} cm^3 s^{-1}, for the electron exchange reaction of O with H which results in de-excitation of the $2p^4$ ^1D term of O to its $2p^4$ ^3P ground term [i.e. the reverse of reaction (8.42)]. Data from [199].*

T (K)	k_r
1000	0.52
2000	0.64
3000	0.71
4000	0.77
5000	0.81
6000	0.84
8000	0.88
10 000	0.91

where the statistical weights, $\omega = (2S+1)(2L+1)$, are $\omega_f = 9$ and $\omega_r = 5$ (the statistical weight of only the O atom need be considered, as that of the H atom remains unchanged).

The possibility that charge transfer might contribute to emission lines from nebulae was investigated initially by Shields *et al.* [200] and subsequently by Clegg and Walsh [201] and by Clegg *et al.* [202]. Clegg and Walsh determined the fraction of neutral hydrogen in the planetary nebulae NGC 7662 and NGC 3918 as 6×10^{-4} and 1×10^{-3} from the relative intensities of O III emission lines observed in these objects.

Certain transitions in the O III spectrum observed in planetary nebulae are excited by a fluorescence process – the 'Bowen' fluorescence mechanism. The fluorescence occurs because of a near-coincidence of the wavelengths of the He II 1s ^2S – 2p $^2P^o$ transition, which is produced subsequent to the radiative recombination of He^{2+},

$$He^{2+} + e^- \rightarrow He^+ + h\nu \tag{8.44}$$

and the O III $2p^2$ $^3P_2$ – 2p 3d $^3P_2^o$ transition, which both fall near 30.4 nm. The He II line excites the O III transition, and the subsequent radiative cascade ('fluorescence') gives rise to transitions in the visible part of the spectrum. The charge transfer process

$$O^{3+} + H \rightarrow O^{2+} + H^+ \tag{8.45}$$

also contributes to populating some of the levels of O^{2+} which intervene in Bowen fluorescence [203].

9
Electron collisions

9.1 Introduction

An elementary calculation of classical mechanics shows that the kinetic energy transferred to a mass M, initially at rest, through the impact of a mass m, initially moving with speed v, is

$$\frac{1}{2}MV'^2 = \frac{1}{2}mv^2 \frac{4mM}{(M+m)^2} \tag{9.1}$$

where V' is the speed of M after the collision (assumed to be elastic). If $m = M$, $\frac{1}{2}MV'^2 = \frac{1}{2}mv^2$, i.e. all of the energy of m is transferred to M. On the other hand, if $m << M$, $\frac{1}{2}MV'^2 \approx \frac{1}{2}mv^2(4m/M)$; only a small fraction, $4m/M$, of the kinetic energy of m is transferred to M. This simple calculation shows that kinetic energy is most efficiently transferred in collisions between particles of comparable or equal mass (e.g. electron–electron, proton–proton) and least efficiently in collisions between a light projectile and a heavy target (e.g. electron–proton). It follows that *electronic* excitation of atoms and molecules (transfer of kinetic energy to the bound electrons) is performed more efficiently by electron than heavy-particle impact.

There is another reason why electron collisions are more effective in exciting electronic degrees of freedom: at a given kinetic temperature, the mean thermal speed of an electron is approximately 43 times greater than that of a proton, and the electron collision frequency is proportionately higher. Furthermore, in the case of collisions with positive ions, the electron collision rate is enhanced by the focusing effect of the coulomb attraction. Thus, when electrons are present in the medium to a significant degree, owing to radiative or collisional ionization of atoms, they tend to dominate the collisional excitation of atoms and ions.

Observations have shown that the process of star formation is characterized by the formation of an accretion disk and accompanied by bipolar jets of matter, which can be traced sometimes to large distances from the protostellar object. These jets are visible owing to their effects on the medium through which they pass, at highly supersonic speeds. The resulting shock waves excite and and dissociate the molecular gas and erode the grain material. Knots of gas are observed – the so-called 'Herbig–Haro' (HH) objects – which emit not only rovibrational transitions of H_2 but also 'forbidden' electronic transitions of atoms, such as C^0 and N^0, and of ions, such as S^+ and Fe^+. It is to these 'forbidden' lines that we now turn.

154   *Electron collisions*

9.2   Selection rules and LS-coupling

9.2.1   *Radiative transitions*

Transitions that are not allowed to electric dipole radiation are termed 'forbidden'. Such transitions dominate the visible spectra of planetary nebulae and 'H II' regions and are also present in the spectra of HH objects, which are associated with the jets from protostars. Although the main discussion of radiative transitions is deferred to the following chapter, it is appropriate to consider here the radiative selection rules, in order to understand the implication of a transition's being 'forbidden'.

The selection rules for electric dipole radiation are: the parity must change; $\Delta S = 0$, where S is the total spin quantum number; $\Delta L = 0, \pm 1$ (but $\Delta L = 0$, when $L = 0$, is not allowed), where L is the total orbital angular momentum quantum number. Strictly speaking, the rules relating to the allowed changes in S and L apply only in the limit of LS-coupling (see below); the more general selection rules are $\Delta J = 0, \pm 1$ (but $\Delta J = 0$, when $J = 0$, is not allowed) and $\Delta M = 0, \pm 1$, where

$$\mathbf{J} = \mathbf{L} + \mathbf{S} \tag{9.2}$$

is the total angular momentum and M is the corresponding projection on the z-axis.

The parity of an electronic state of an atom or an ion is given by

$$p = (-1)^{\sum_i l_i} \tag{9.3}$$

where l_i is the orbital angular momentum quantum number of electron i, and the summation extends over all the electrons of the atom. The Pauli exclusion principle dictates that there is always an even number of electrons in a closed shell (determined by a given value of the principal quantum number, n) and in a closed sub-shell (determined by a given value of the orbital angular momentum quantum number, l), as the electrons are paired, with 'up' and 'down' spin orientations. Thus, $\sum_i l_i$ is always an even integer for closed shells and sub-shells, and the parity is determined by the *valence* electrons in the open sub-shells. Thus, in the 'ground' configuration of atomic carbon ($1s^2 2s^2 2p^2$), which has six electrons, the parity is determined by the $2p^2$ electrons and hence is even ($p = +1$). On the other hand, in the ground configuration of atomic nitrogen ($1s^2 2s^2 2p^3$), which has seven electrons, the parity is odd ($p = -1$).

In light atoms, it is a good approximation to vector-couple separately the orbital and spin angular momenta of the individual electrons:

$$\mathbf{L} = \sum_i \mathbf{l}_i \tag{9.4}$$

$$\mathbf{S} = \sum_i \mathbf{s}_i \tag{9.5}$$

Applying the Pauli principle, which forbids two electrons from having the same set of quantum numbers, to this 'LS-coupling' scheme, yields the 'terms' which are allowed, for a given electronic configuration. The terms are denoted by the spin multiplicity, $2S + 1$ (the number of possible values of M_S), and by L, in the form ^{2S+1}L (and with $L = 0, 1, 2, 3, 4, \ldots$ being replaced by the spectroscopic notation S, P, D, F, G, …). In the ground configuration of atomic carbon, there exist three terms, ^{3}P, ^{1}D and ^{1}S. The total multiplicity of the ground

9.2 Selection rules and LS-coupling

configuration, $\prod_{i=1}^{2}(2l_i+1)(2s_i+1) = 36$, where the product extends over the two electrons in the incomplete $2p^2$ sub-shell. Equations (9.4) and (9.5) would allow $L = 0, 1, 2$ and $S = 0, 1$, namely six terms with a total multiplicity of $\sum_{j=1}^{6}(2L_j+1)(2S_j+1) = 36$, once again; but the terms ^{1}P, ^{3}D and ^{3}S are excluded by the Pauli principle.

The electromagnetic spin–orbit interaction gives rise to a splitting of the term energies into levels corresponding to different values of J, the total electronic angular momentum quantum number. The ^{3}P term comprises three levels, ^{3}P$_0$, ^{3}P$_1$, ^{3}P$_2$, corresponding to the three values of J, determined by $|L-S| \leq J \leq L+S$. The separations of these *fine structure levels* are small compared with the separations of the terms. In atomic carbon, for example, the separations of the fine structure levels of the ^{3}P ground term are approximately 20 cm^{-1}, whereas the terms of the ground configuration are separated by approximately 10 000 cm^{-1}. As all terms within a given configuration have the same parity [by its definition, equation (9.3)], transitions between the terms are forbidden to electric dipole radiation (which requires a change in parity). Such transitions may additionally violate the approximate selection rules on ΔL and ΔS. In the example of atomic carbon, the transitions ^{3}P–^{1}D and ^{3}P–^{1}S are forbidden by the spin-change selection rule, and the ^{1}D–^{1}S transition violates the selection rule for ΔL. Such 'forbidden' transitions can still occur, owing to higher order terms (electric quadrupole, magnetic dipole, ...) in the multipole expansion of the radiation field; but their transition probabilities are small.

9.2.2 Collisional transitions

The selection rules considered above apply to radiative transitions. Transitions induced by collisions, on the other hand, involve the perturber (to be denoted by '2'), as well as the target atom or ion ('1'), and the restrictions are determined by the properties of the system of perturber-plus-target. For example, the total parity, p, includes the contributions of the perturber, the target, and the relative angular momentum, l, of the perturber–target system:

$$p = (-1)^l p_1 p_2 \tag{9.6}$$

The total parity is conserved in the collision; but the parity of the target, p_1, may either change or remain unchanged, subject to equation (9.6) being satisfied. Similarly, the total angular momentum

$$\mathbf{J} = \mathbf{l} + \mathbf{J_1} + \mathbf{J_2} \tag{9.7}$$

must be conserved. In the limit of *LS*-coupling, the total orbital angular momentum and the total spin are conserved separately. The total spin is the sum of the contributions of the target and projectile:

$$\mathbf{S} = \mathbf{S_1} + \mathbf{S_2} \tag{9.8}$$

If $S_2 = 0$, which is the case, for example, of the important perturbers He and H$_2$ in their electronic ground states, then $S = S_1$, the spin angular momentum of the target. Thus, collisions with ground state He or H$_2$ cannot change the value of the spin quantum number, S_1, of the target. It follows that, in the example of atomic carbon, transitions ^{3}P $\rightarrow$ ^{1}D and

$^3P \to {}^1S$ do not occur in such collisions. On the other hand, if $S_2 = \frac{1}{2}$, which is the case of a free electron or a hydrogen atom then, for example,

$$\left|S_1 - \frac{1}{2}\right| \leq S \leq S_1 + \frac{1}{2}$$

Transitions between target states with $S_1 = 1$ and $S_1 = 0$, as in the example, are then possible, via the (common) state of total spin angular momentum $S = \frac{1}{2}$.

The process involved – electron exchange between the perturber and the target – occurs more readily with a free electron than with an electron bound to the proton of a hydrogen atom. Consequently, when transitions such as [C I] $^1D_2 \to {}^3P_1$ and $^1D_2 \to {}^3P_2$ are observed, at 0.983 and 0.985 µm, respectively, the emitting gas must be at least partially ionized. (In this example, we have used the spectroscopic notation of Roman numerals to indicate the spectrum, with I denoting a neutral atom, II a singly ionized ion, etc. The square brackets indicate transitions that are forbidden to electric dipole radiation.)

9.3 Electron collisional excitation

The excitation of atoms and ions in collisions with electrons is a process that has been studied extensively. The results that were obtained by the 'Opacity' and 'Iron' projects [204, 205] are the most comprehensive collections of such data that are currently available, providing benchmarks for subsequent calculations. Computations of electron collision cross-sections and rate coefficients are similar in some respects to problems in heavy particle scattering, considered in earlier chapters. However, there exist also important differences, to which we refer below.

Collisions between electrons and atoms or ions involve only electronic degrees of freedom and the corresponding electron wave functions. The nucleus contributes to the coulomb field and possibly to the total spin, but the rotational and vibrational motions of the nuclei, which occur in collisions with *molecules*, are absent in collisions with atoms and ions. Hence, there is no necessity for approximations of the Born–Oppenheimer type (cf. Section 4.2), and electron–atom collisions are, in principle, simpler to treat and more amenable to 'exact' solutions.

The coulomb interaction between an incoming electron (perturber) and a one-electron atom or ion (target) may be expressed exactly as

$$V = -\frac{Z}{r_2} + \frac{1}{r_{12}} \qquad (9.9)$$

where Z is the atomic number of the nucleus of the target, r_2 is the separation of the perturber electron from the nucleus, and r_{12} is the separation of the incoming electron from the electron of the target; atomic units are used. The target is assumed to possess only one electron, in order to simplify the presentation; but the analysis may be generalized to many-electron atoms and ions.

The term in equation (9.9) which represents the coulomb repulsion between the two electrons may be expanded using the identity

$$\frac{1}{r_{12}} = \sum_{\lambda=0}^{\infty} v_\lambda(r_1, r_2) P_\lambda(\hat{\mathbf{r}}_1 \cdot \hat{\mathbf{r}}_2) \qquad (9.10)$$

9.3 Electron collisional excitation

where

$$v_\lambda(r_1, r_2) = \frac{r_<^\lambda}{r_>^{\lambda+1}} \tag{9.11}$$

where $r_<$ is the lesser and $r_>$ is the greater of r_1 and r_2, the distances of the target electron and of the perturber electron, respectively, from the nucleus. The form of the expansion (9.10) is similar mathematically to equation (4.39), introduced in the discussion of atom–diatom collisions. If the two electrons are considered to be formally distinguishable, Schrödinger's equation may be reduced to a set of coupled, ordinary differential equations which are analogous to those derived in Chapter 4 [equation (4.57)]:

$$\left[\frac{d^2}{dr_2^2} - \frac{l_2(l_2+1)}{r_2^2} + \frac{2Z}{r_2} + k_{nl_1l_2}^2\right] F(nl_1l_2L|r_2)$$

$$= 2 \sum_{n'l'_1l'_2} \sum_{\lambda=0}^\infty y_\lambda(P_{nl_1}P_{n'l'_1}|r_2) f_\lambda(l_1l_2, l'_1l'_2; L) F(n'l'_1l'_2L|r_2) \tag{9.12}$$

In (9.12), atomic units are used; nl_1 are the quantum numbers of the bound electron of the target atom or ion, and $\mathbf{L} = \mathbf{l_1} + \mathbf{l_2}$ is the total orbital angular momentum. The radial functions of the target and perturber electrons are denoted $P(nl_1|r_1)$ and $F(nl_1l_2L|r_2)$, respectively, and f_λ is a Percival–Seaton coefficient (cf. equation 4.56). The functions y_λ are defined by

$$y_\lambda(P_{nl_1}P_{n'l'_1}|r_2) = \frac{1}{r_2^{\lambda+1}} \int_0^{r_2} P^*(nl_1|r_1) P(n'l'_1|r_1) r_1^\lambda dr_1$$

$$+ r_2^\lambda \int_{r_2}^\infty P^*(nl_1|r_1) P(n'l'_1|r_1) \frac{1}{r_1^{\lambda+1}} dr_1 \tag{9.13}$$

The similarity of equations (9.12) and (4.57) should be evident.

The analysis so far assumes the electrons to be distinguishable and neglects their spin, which is treated as a 'spectator' quantum number. In fact, the electrons are identical fermions, and so the total electronic wave function (of perturber-plus-target) must be asymmetric under electron exchange, in order to satisfy the Pauli exclusion principle. After anti-symmetrization of the wave function, the scattering equations become

$$\left[\frac{d^2}{dr_2^2} - \frac{l_2(l_2+1)}{r_2^2} + \frac{2Z}{r_2} + k_{nl_1l_2}^2\right] F(nl_1l_2LS|r_2)$$

$$= 2 \sum_{n'l'_1l'_2} \sum_{\lambda=0}^\infty y_\lambda(P_{nl_1}P_{n'l'_1}|r_2) f_\lambda(l_1l_2, l'_1l'_2; L) F(n'l'_1l'_2LS|r_2)$$

$$+ (-1)^S \left[\delta_{\lambda 0}(E_n + E_{n'} - E)\Delta(P_{nl_1}F_{n'l'_1l'_2LS}) + y_\lambda(P_{nl_1}F_{n'l'_1l'_2LS}|r_2)\right]$$

$$\times g_\lambda(l_1l_2, l'_1l'_2; L) P(n'l'_1|r_2) \tag{9.14}$$

158   *Electron collisions*

In equation (9.14), E is the total energy and E_n, $E_{n'}$ are eigenenergies of the one-electron atom or ion (n is the principal quantum number). In addition,

$$\Delta(P_{nl_1}F_{n'l'_1 l'_2 LS}) = \int_0^\infty P^*(nl_1|r)F(n'l'_1 l'_2 LS|r)\mathrm{d}r \tag{9.15}$$

and

$$g_\lambda(l_1 l_2, l'_1 l'_2; L) = (-1)^{l_1+l_2-L} f_\lambda(l_1 l_2, l'_2 l'_1; L) \tag{9.16}$$

were defined by Percival and Seaton [46]. The total spin quantum number, $S = 0$ or $S = 1$; both S and L, the total orbital angular momentum, are conserved in the limit of LS-coupling. The total parity of the wave function, $(-1)^{l_1+l_2} = (-1)^{l'_1+l'_2}$, is conserved rigorously. The additional terms in equation (9.14), as compared with equation (9.12), involve integrals of the scattering wave function, $F_{n'l'_1 l'_2 LS}$, which transform the differential equations into *integro*-differential equations. Sophisticated numerical methods have been developed to solve equations of this kind.

The treatment of many-electron targets is more complex but retains the same basic features as the scattering of an electron by a one-electron atom or ion.

9.4   Resonances

Two kinds of resonance need to be considered in scattering events, the so-called 'shape' and 'Feshbach' resonances. Shape resonances tend to be more important in heavy-particle scattering, and Feshbach resonances dominate in electron collisions.

9.4.1   Shape resonances

Associated with any collision event is the 'centrifugal' potential, $l(l+1)/(2\mu r^2)$, where l is the relative orbital angular momentum quantum number and μ is the reduced mass of the perturber-plus-target system; $\mu = m_1 m_2/(m_1 + m_2)$, where m_1 and m_2 are the masses of the target and of the perturber. In electron–atom scattering, $\mu \approx m_e = 1$ when atomic units are used. The centrifugal term appears on the left-hand sides of equations (4.57) and (9.14) above. We consider cases where the interaction potential $V(r)$ is attractive and $V(r) \sim -r^{-n}$, with $n > 2$. An important example of such an interaction is the polarization potential between a neutral target and a charged perturber. In this case, $n = 4$ and $V(r) = -\alpha/(2r^4)$ (in atomic units), where α is the polarizability of the neutral, and hence the effective potential is

$$V_{\mathrm{eff}}(r) = -\frac{\alpha}{2r^4} + \frac{l(l+1)}{2\mu r^2} \tag{9.17}$$

The effective potential is dominated by the (repulsive) centrifugal term at long range and has a maximum at the point where $\mathrm{d}V_{\mathrm{eff}}(r)/\mathrm{d}r = 0$. As already seen in Chapter 6, equation (9.17) may be written in terms of the classical impact parameter b and the collision energy E as

$$V_{\mathrm{eff}}(r) = -\frac{\alpha}{2r^4} + \frac{Eb^2}{r^2} \tag{9.18}$$

and the maximum in the effective potential occurs at $r = (\alpha/E)^{\frac{1}{2}}/b$, at which point $V_{\mathrm{eff}} = E^2 b^4/(2\alpha)$. When this maximum value of V_{eff} is equal to the total energy, E, the

9.4 Resonances

radial kinetic energy of the perturber and target is zero: they perform orbital motion at a point of unstable equilibrium. The classical expression for the cross-section for orbiting is

$$\sigma = \pi b^2 = \pi \left(\frac{2\alpha}{E}\right)^{\frac{1}{2}} \qquad (9.19)$$

and the corresponding rate coefficient is

$$\langle \sigma v \rangle = 2\pi \left(\frac{\alpha}{\mu}\right)^{\frac{1}{2}} \qquad (9.20)$$

In the example of proton collisions with hydrogen atoms, $\alpha = 4.5$ and $\mu = 918$ in atomic units, whence $\langle \sigma v \rangle = 2.7 \times 10^{-9}$ cm^3 s^{-1}. For electron collisions with hydrogen atoms, on the other hand, $\mu = 1$ and $\langle \sigma v \rangle = 8.2 \times 10^{-8}$ cm^3 s^{-1}.

To the value of the impact parameter, b, for orbiting at the collision energy, E, there corresponds a value of the relative orbital angular momentum quantum number, l, for which there is an enhancement of the cross-section; this is the condition for a 'shape' resonance.

9.4.2 Feshbach resonances

In collisions of electrons with atoms or ions, resonant excitation of the target can occur. This process may be viewed as excitation of the target electron and simultaneous capture of the perturbing electron into a doubly-excited state of the perturber-plus-target system, followed by relaxation of the target to a singly-excited state in the course of the same collision event. The mechanism is illustrated by means of an example in Fig. 9.1.

Let us suppose that the target is a positive ion with nuclear charge Z and with $N = Z - 1$ electrons. Converging on the ground state of this ion is a Rydberg series of singly-excited

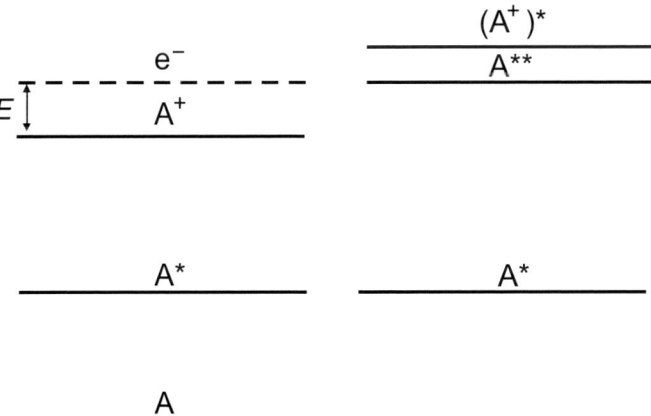

Figure 9.1 Illustrating the resonant capture by a target ion A^+, initially in its ground energy level, of an incident electron e^-. The energy of the incident electron lies in the first ionization continuum of the atom A, i.e. the incident electron has a positive total energy E relative to the ground state of the ion A^+. When the ion is excited by the incident electron to the level $(A^+)^*$, the energy of the incident electron may coincide with that of a doubly-excited state A^{**} of the atom. The doubly-excited state A^{**} eventually decays to A^+ + e^-. This process, of temporary capture of the incident electron by the excited ion $(A^+)^*$, manifests itself as a resonance in the electron scattering cross-section.

states of the corresponding atom with $N + 1 = Z$ electrons. Furthermore, to each excited state of the N-electron ion converges the corresponding Rydberg series. If the energy of the incident electron coincides with the energy of one of the Rydberg levels, it can become bound in a doubly-excited state of the $(N + 1)$-electron atom. The symmetry of the doubly-excited state, i.e. the values of the total orbital and spin angular momenta and parity, must be the same as the overall symmetry of the incident electron and target ion. The doubly-excited states of the atom can decay and hence stabilize by the emission of radiation, in which case *dielectronic recombination* has occurred; but the more probable process is radiationless relaxation, either back to the ground state of the ion [plus a continuum (free) electron, whose kinetic energy is equal to the initial kinetic energy; in this case, *elastic scattering* of the electron has occurred], or into an excited state of the ion [plus a free electron, whose kinetic energy is lower than initially by an amount equal to the excitation energy of the ion; in this case, *inelastic scattering* of the electron has occurred].

The width, ΔE, of a resonance is related to its lifetime, τ, through the uncertainty principle

$$\Delta E \, \tau \approx h$$

The shorter the lifetime, the broader the resonance. Resonances occurring in the $3d^6\,4s\,^6D - 3d^6\,4s\,^4D$ transition of Fe^+, via the ^5G state (i.e. $L=4$, $S=2$, even parity) of the $(Fe^+ + e^-)$ system, are illustrated in Fig. 9.2. The resonance structure is seen to be complex and to dominate the contributions to the collision strength in the energy region in which resonances occur.

The *collision strength*, Ω, was introduced in Chapter 4. It is defined through

$$\pi \Omega_{i,j} = \sigma(i \leftarrow j) k_j^2 \omega_j \tag{9.21}$$

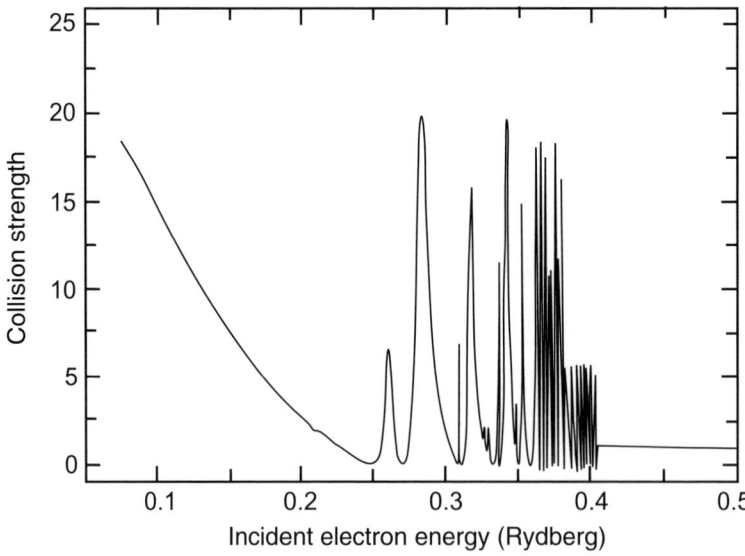

Figure 9.2 Resonances in the $3d^6\,4s\,^6D - 3d^6\,4s\,^4D$ transition of Fe^+, via the ^5G state (i.e. $L=4$, $S=2$, even parity) of the $(Fe^+ + e^-)$ system [206].

where ω_j is a statistical weight and $k_j^2 = 2E_j$, in atomic units; E_j is the electron collision energy, relative to state j. From the equation of detailed balance, which relates the cross-sections, $\sigma(j \leftarrow i)$ and $\sigma(i \leftarrow j)$, for forward and reverse transitions,

$$\sigma(j \leftarrow i)\omega_i E_i = \sigma(i \leftarrow j)\omega_j E_j \tag{9.22}$$

we see that the collision strength is symmetric in the initial and final states, i and j; it is also dimensionless. The rate coefficient at kinetic temperature T for the transition $i \leftarrow j$, induced by electron collisions, may be written in terms of an integral over the collision strength:

$$\langle \sigma v \rangle_{i \leftarrow j} = \frac{8.63 \times 10^{-6}}{\omega_j T^{\frac{1}{2}}} \int_0^\infty \Omega_{i,j}(x_j) e^{-x_j} \, dx_j \tag{9.23}$$

When resonances are present, $\Omega_{i,j}(x_j)$ varies rapidly with $x_j \equiv E_j/(k_B T)$; the integral in equation (9.23) must then be evaluated with due care. The results of such calculations are often presented in the form of tabulations of the integral

$$\Upsilon_{i,j} = \int_0^\infty \Omega_{i,j}(x_j) e^{-x_j} \, dx_j$$

as a function of the kinetic temperature, T.

9.5 Forbidden line emission from Herbig–Haro objects

We have already mentioned, in Section 9.1, that the spectra of Herbig–Haro objects – knots of gas present in the jets emanating from proto-stars – contain not only rovibrational transitions of H_2 but also forbidden lines of atoms and ions, including C, N, S^+ and Fe^+. The observed transitions of [C I] $^1D - {}^3P$ and of [Fe II] $a^4D - a^6D$ involve a change of the electron spin. Arguments based on conservation laws, in LS-coupling, show that collisions with free electrons are likely to dominate the excitation of these transitions.

The perturbers which are sufficiently abundant and could, in principle, excite the forbidden lines are H_2, He, H and e^-. The ground electronic state of H_2, $X^1\Sigma_g^+$, is the only state to be significantly populated, and it is a singlet state ($S_2 = 0$). In the case of [Fe II] $a^4D - a^6D$, the electron spin changes from $S_1 = \frac{3}{2}$ to $S_1 = \frac{5}{2}$ in the transition. If H_2 remains in its electronic ground state, such a transition cannot occur (in LS-coupling) as the total spin of the perturber-plus-target would not be conserved. An analogous argument also excludes He. On the other hand, if the perturber is a hydrogen atom or a free electron, the transition can occur via a common state of the perturber-plus-target with total spin $S = 2$.

The spin-changing process described above proceeds by the exchange of an electron between the perturber (2) and the target (1). In the case of collisions of Fe^+ with atomic hydrogen, this involves exchanging electrons that are initially in the ground states of the target ion and the perturber atom. The ionization potentials of H (13.60 eV) and of Fe^+ (16.16 eV) are large and differ substantially from each other, i.e. the process is far from being 'resonant'. It follows that spin-changing transitions of the ion are unlikely to be induced by atomic hydrogen; they are much more likely to be excited by collisions with free electrons.

The above argument assumes that LS-coupling is valid, i.e. that the orbital and spin angular momenta are conserved separately. In fact, because of departures from LS-coupling, only the total angular momentum,

$$\mathbf{J} = \mathbf{L} + \mathbf{S}$$

is an exactly conserved angular momentum. The extent of departures from an LS-coupling scheme is reflected in the relative magnitudes of the separations between fine structure levels of a given term, and of the separations of the terms. In the example of [Fe II] $a^4D - a^6D$, the separation of adjacent fine structure levels of the a^4D term or the a^6D term is of the order of 200 cm^{-1}, whereas the separation of the terms is about 8000 cm^{-1}. Thus, the effects of the departures from LS-coupling are expected to be small. These departures arise from magnetic interactions and are relativistic in origin. Relativistic effects become important for high stages of ionization and for heavy elements: a large nuclear charge results in electrons acquiring relativistic speeds in the course of their orbital motion.

We may conclude that the forbidden lines observed in Herbig–Haro objects are excited principally by electron collisions. It follows that the analysis of their intensities can provide information on the free electron density, and hence the degree of ionization, in these objects.

10

Photon collisions

10.1 Introduction

Photons carry energy ($h\nu$) and momentum (h/λ), where ν is the frequency and λ is the wavelength of the photon; both the energy and the momentum can be transferred to other particles. In these respects, the scattering of photons from atoms is analogous to electron–atom scattering. However, there are important differences between these two categories of collision process: the photon is neutral, whereas the electron is negatively charged; the wavelength of a photon is much larger than the de Broglie wavelength of an electron with the same energy, in the range of energies that concerns us here. It follows that the momentum of the photon is much less than the momentum of the electron; this may be seen as follows.

Consider a photon with an energy of 1 eV, corresponding to a wavelength of approximately 1 μm. The de Broglie wavelength, λ_e, of an electron with the same energy (1 eV) is approximately 1 nm, i.e. 10^3 times smaller. More generally, the ratio of the wavelengths of a photon and an electron with the same (non-relativistic) energy, E (eV), is given by

$$\frac{\lambda}{\lambda_e} = \frac{1.011 \times 10^3}{[E(\text{eV})]^{\frac{1}{2}}}$$

This expression is valid for $E \ll m_e c^2 = 0.51$ MeV, where m_e is the rest mass of the electron. For transitions at ultraviolet or longer wavelengths, $E \lesssim 10$ eV, the photon wavelength $\lambda \gg \lambda_e$, the de Broglie wavelength of the electron with the same energy. Furthermore, whilst the wavelength of the electron is comparable with atomic dimensions, the wavelength of the corresponding photon is much larger than atomic dimensions. It follows that quantum mechanics are essential to treat electron–atom collisions, whereas a *classical* treatment of the radiation field is generally adequate for the treatment of photon–atom interactions.

10.2 The oscillator strength

Let us consider a gas containing n absorbing atoms per unit volume. The rate of decrease in the intensity of radiation of frequency ν, incident in the x-direction, owing to absorption by the atoms, is given by

$$\frac{dI_\nu}{dx} = -\kappa_\nu I_\nu \qquad (10.1)$$

where κ_ν is the opacity (also known as the absorption coefficient). According to classical theory, if the atom comprises f harmonic oscillators (oscillating dipoles), then

$$\kappa_\nu = n \frac{\pi e^2}{m_e c} f \phi_\nu \tag{10.2}$$

where e is the electron charge, m_e is the electron mass, c is the speed of light, and

$$\phi_\nu = \frac{\gamma}{4\pi^2} \frac{1}{(\nu - \nu_0)^2 + (\gamma/4\pi)^2} \tag{10.3}$$

is the 'natural' or 'Lorentz' line profile function. In equation (10.3), ν_0 is the central frequency of the absorption line and γ is its full width at half minimum (FWHM) intensity. The profile function, ϕ_ν, is normalized such that

$$\int \phi_\nu d\nu = 1 \tag{10.4}$$

where the integration extends across the entire absorption profile.

In the quantum mechanical formulation, absorption occurs in a specific spectral line and is associated with a transition from a lower level i to a higher level j. As may be seen from the analysis in Section 7.2, the corresponding expression for the opacity is

$$\kappa_\nu = \frac{h\nu}{c} [B(2 \leftarrow 1)n_1 - B(1 \leftarrow 2)n_2] \phi_\nu \tag{10.5}$$

where

$$B(2 \leftarrow 1)\omega_1 = B(1 \leftarrow 2)\omega_2 \tag{10.6}$$

relates the Einstein B-coefficients and

$$B(1 \leftarrow 2) = A(1 \leftarrow 2) \frac{c^3}{8\pi h \nu^3} \tag{10.7}$$

relates the B- and A-coefficients; ω_1 and ω_2 are the statistical weights (degeneracies) of the energy levels.

The first term on the right-hand side of equation (10.5) corresponds to absorption from the lower level 1 and the second term to stimulated emission from the upper level 2. We define the *oscillator strength* for absorption from the lower level 1 to the upper level 2, f_{12}, and the oscillator strength for emission from 2 to 1, f_{21}, such that

$$\frac{\pi e^2}{m_e c} (n_1 f_{12} + n_2 f_{21}) \phi_\nu \equiv \frac{h\nu}{c} [B(2 \leftarrow 1)n_1 - B(1 \leftarrow 2)n_2] \phi_\nu \tag{10.8}$$

is the corresponding identity for the opacity. Then, using equations (10.6) and (10.7), we obtain

$$f_{12} = \frac{m_e c}{8\pi^2 e^2} \lambda^2 A(1 \leftarrow 2) \frac{\omega_2}{\omega_1} \tag{10.9}$$

for the oscillator strength for absorption, and

$$f_{21} = -\frac{\omega_1}{\omega_2} f_{12} \qquad (10.10)$$

for the oscillator strength for emission. Thus, the oscillator strength for absorption, f_{12}, is positive, whereas that for emission, f_{21}, is negative.

It will be recalled that, classically, f is the number of oscillators, i.e. the number of optically active electrons in the atom. To this classical definition there corresponds the quantum mechanical *f-sum rule*

$$\sum_1 f_{21} + \sum_3 f_{23} = n \qquad (10.11)$$

where n is the number of optically active electrons. The first summation in equation (10.11) extends over all levels '1' lower than level 2, and the second summation over all levels '3' higher than level 2, including the continuum; the contribution of the first summation is negative. If level 2 is the ground state, the first summation in equation (10.11) vanishes.

10.3 The transition probability

The quantum theory of radiation enables the A-coefficient to be expressed in terms of the *line strength*; specifically

$$A(i \leftarrow j) = \frac{64\pi^4 \nu^3}{3hc^3} S_{ij} \qquad (10.12)$$

where

$$S_{ij} = \sum_s |\langle i| - e\mathbf{r}_s |j\rangle|^2 \qquad (10.13)$$

and $\mathbf{r}_s$ is the position vector of the active electron s, relative to the nucleus. Thus, the line strength is the square of the electric dipole matrix element and may be expanded as

$$S_{ij} = e^2 \sum_s \left[|\langle i|x_s|j\rangle|^2 + |\langle i|y_s|j\rangle|^2 + |\langle i|z_s|j\rangle|^2 \right] \qquad (10.14)$$

when $\mathbf{r}_s = (x_s, y_s, z_s)$ is expressed in cartesian coordinates. We note that the line strength is symmetric, i.e. $S_{ji} = S_{ij}$.

In general, the energy levels involved in a transition are degenerate: they each comprise several quantum states of the same energy. Suppose that there are ω_m degenerate states i belonging to the energy level m and ω_n degenerate states j belonging to the energy level n. Then, the spontaneous transition probability linking the upper level n to the lower level m is obtained by summing equation (10.12) over the degenerate final (lower) states and averaging over the degenerate initial (upper) states:

$$A(m \leftarrow n) = \frac{1}{\omega_n} \sum_{i,j} A(i \leftarrow j) \qquad (10.15)$$

166   *Photon collisions*

The lifetime of the upper level is

$$\tau_n = \frac{1}{\sum_m A(m \leftarrow n)} \quad (10.16)$$

where the summation extends over all lower levels m.

In the case of an electric quadrupole transition, the expression for the transition probability is

$$A(m \leftarrow n) = \frac{1}{\omega_n} \frac{32\pi^6 \nu^5}{5hc^5} \sum_{i,j} S_{ij}^{(q)} \quad (10.17)$$

where $S_{ij}^{(q)}$ is the square of the electric quadrupole matrix element. We note from equations (10.12) and (10.17) that the probability of an electric dipole transition is proportional to ν^3, whereas the probability of an electric quadrupole transition is proportional to ν^5. The expression for the probability of a magnetic dipole transition is analogous to equation (10.12), with S_{ij} replaced by $S_{ij}^{(m)}$ on the right-hand side, where $S_{ij}^{(m)}$ is the square of the magnetic dipole matrix element.

The line profile function (10.3) (the 'natural' or 'Lorentz' line profile) is not appropriate when the broadening of the spectral line arises from random thermal or turbulent motions. In this case, the normalized line profile takes the Doppler form

$$\phi_\nu = \frac{1}{\nu_0} \left(\frac{\beta}{\pi}\right)^{\frac{1}{2}} \exp\left[-\beta\left(\frac{\nu - \nu_0}{\nu_0}\right)^2\right] \quad (10.18)$$

where $\beta = mc^2/(2k_B T_D)$; m is the mass of the emitting atom and T_D is the Doppler temperature. When both natural and Doppler broadening contribute significantly to the line profile, (10.3) and (10.18) must be convolved, yielding a *Voigt profile* function

$$\phi_\nu = \frac{1}{\nu_0} \left(\frac{\beta}{\pi}\right)^{\frac{1}{2}} H(a,x) \quad (10.19)$$

In equation (10.19), $x \equiv \beta^{\frac{1}{2}}(\nu - \nu_0)/\nu_0$, $a \equiv \beta^{\frac{1}{2}} \gamma/(4\pi \nu_0)$ and

$$H(a,x) = \frac{a}{\pi} \int_{-\infty}^{\infty} \frac{e^{-t^2}}{a^2 + (t-x)^2} dt \quad (10.20)$$

is the Voigt function.

10.4   Photoionization and radiative recombination

In regions of the interstellar medium which are permeated by an ultraviolet radiation field, atoms and molecules can be photoionized. In the immediate vicinities of hot stars, which emit photons with energies up to and beyond the H I Lyman limit (13.6 eV), hydrogen is predominantly in the form of H^+, and the fractional ionization, n_e/n_H, is close to unity. Thus, in 'H II' regions and planetary nebulae, electron collisions largely determine the emission line spectrum. Electron scattering from protons gives rise to the 'free-free' (*bremsstrahlung*) continuum emission, which is observed at radio wavelengths. The radiative recombination

10.4 Photoionization and radiative recombination

of electrons with positive ions, principally protons, also gives rise to continuum emission, and, as a consequence of the subsequent radiative cascade, to recombination lines. Radiative recombination is the reverse of photoionization, and consequently the rates of these two processes are related, as we shall see below. The discussion will be directed towards the photoionization of atomic hydrogen and the radiative recombination of protons with electrons; these (one-electron) processes are important, in view of the high cosmic abundance of hydrogen.

The cross-section relating to the photoionization of atomic hydrogen in the energy level with the principal quantum number n, and notably in its ground state, $n = 1$, has been known essentially exactly since the work of H. A. Kramers and of J. A. Gaunt in the 1920s. Generalizing to a one-electron atom with atomic number Z (i.e. nuclear charge Ze), the photoionization cross-section may be written in the form

$$a_\nu(Z, n) = \frac{8h^3}{3^{\frac{3}{2}}\pi^2 e^2 m_e^2 c} \frac{n}{Z^2} \left(\frac{\nu_n}{\nu}\right)^3 g_\nu \tag{10.21}$$

where ν is the frequency of the photon, ν_n is the threshold ionization frequency, and g_ν is the Kramers–Gaunt g-factor, which is of order unity; $\nu_n = \nu_1 Z^2/n^2$, where ν_1 is the threshold ionization frequency of atomic hydrogen in its ground state. Substituting the numerical values of the constants in equation (10.21), we obtain

$$a_\nu(Z, n) = 7.91 \times 10^{-18} \frac{n}{Z^2} \left(\frac{\nu_n}{\nu}\right)^3 g_\nu(Z, n) \tag{10.22}$$

when a_ν is expressed in units of cm^2.

The g-factor depends on the principal quantum number, n, and the energy of the photoelectron, expressed relative to the ionization threshold energy, i.e. on $(\nu - \nu_n)/\nu_n$. For atomic hydrogen ($Z = 1$) in its ground state ($n = 1$), the photoionization cross-section is

$$a_{\nu_1} = 6.30 \times 10^{-18} \text{ cm}^2$$

at threshold ($\nu = \nu_1$); thus, $g_{\nu_1} = 0.80$. Furthermore, $g_{\nu_n} \to 1$ as n increases. Thus, an error in a_ν of no more than 20% is made by setting $g_\nu = 1$.

The radiative recombination of an ion A$^+$ with an electron e$^-$, into energy level n of the product atom, A,

$$A^+ + e^- \to A(n) + h\nu \tag{10.23}$$

is the inverse of the photoionization of A in level n; the rates of these processes are consequently related. As we shall now see, the relationship between the photoionization cross-section, a_ν, and the recombination rate coefficient, α_ν, may be derived by considering the limiting case of thermodynamic equilibrium, when the rates (per unit volume of gas) of reaction (10.23) and its inverse are equal and the Saha relation also applies.

In thermodynamic equilibrium, the energy density of radiation in the frequency interval $\nu \to \nu + d\nu$ is given by a Planck distribution at temperature T, namely

$$u_\nu = \frac{8\pi h \nu^3}{c^3} \frac{1}{\exp[(h\nu)/(k_B T)] - 1} \tag{10.24}$$

Photon collisions

The rate per unit volume of photoionization of atom A in level n by photons of frequency ν is $n[A(n)]a_\nu(n)c/(h\nu)u_\nu d\nu$, where $n[A(n)]$ is the number density of atoms in state n.

Let us now consider recombination to level n of an electron with speed in the range $v \to v + dv$; the kinetic energy of the electron is $m_e v^2/2 = h(\nu - \nu_n)$, where ν_n is the photoionization threshold frequency. If the cross-section for this recombination process is denoted $\sigma_v(n)$, the rate per unit volume of recombination is $n(A^+)n_e[\sigma_v(n) + Cu_\nu]vf(v,T)dv$, where the second term in parentheses represents the contribution of *stimulated* recombination and is proportional to the radiation density, u_ν; C will be determined below. In thermodynamic equilibrium, the velocity distribution of the electrons is Maxwellian:

$$f(v,T) = 4\pi \left(\frac{m_e}{2\pi k_B T}\right)^{\frac{3}{2}} v^2 \exp\left(-\frac{m_e v^2}{2k_B T}\right) \tag{10.25}$$

Equating the rates per unit volume of recombination and ionization, we obtain

$$n(A^+)n_e[\sigma_v(n) + Cu_\nu]vf(v,T)dv = n[A(n)]a_\nu(n)\frac{c}{h\nu}u_\nu d\nu \tag{10.26}$$

In thermodynamic equilibrium, the Saha relation applies, whence

$$\frac{n[A(n)]}{n(A^+)n_e} = \frac{\omega_n}{2\omega_+}\left(\frac{h^2}{2\pi m_e k_B T}\right)^{\frac{3}{2}} \exp\left(\frac{h\nu_n}{k_B T}\right) \tag{10.27}$$

where ω_n is the degeneracy (statistical weight) of level n of the atom A, and ω_+ is the degeneracy of the ion A^+ ($\omega_n = 2n^2$ and $\omega_+ = 1$ when A is a one-electron atom); the (spin) degeneracy of the free electron is $2s + 1 = 2$. Using equations (10.24), (10.25) and (10.27), equation (10.26) becomes

$$\left[\exp\left(\frac{h\nu}{k_B T}\right) - 1\right]\sigma_v(n) + \frac{8\pi h\nu^3}{c^3}C$$
$$= \frac{\omega_n}{\omega_+}\frac{h\nu}{m_e c^2}\frac{h\nu}{m_e v^2}a_\nu(n)\exp\left(\frac{h\nu}{k_B T}\right) \tag{10.28}$$

where we have made use of the relations $h\nu = h\nu_n + m_e v^2/2$ and $h d\nu = m_e v dv$. Equation (10.28) is an identity which is valid for all temperatures, T, and therefore

$$\sigma_v(n) = \frac{\omega_n}{\omega_+}\frac{h\nu}{m_e c^2}\frac{h\nu}{m_e v^2}a_\nu(n) \tag{10.29}$$

and

$$C = \frac{c^3}{8\pi h\nu^3}\sigma_v(n) \tag{10.30}$$

These relations apply whether thermodynamic equilibrium has been attained or not. However, the Saha relation (10.27) applies *only* in thermodynamic equilibrium. Under more general conditions, the rate of radiative recombination is given by

$$n(A^+)n_e\sigma_v(n)\left[1 + \frac{W}{\exp[(h\nu)/(k_B T_{bb})] - 1}\right]vf(v,T_e)dv$$
$$\equiv n(A^+)n_e\alpha_v(n,T_e)dv \tag{10.31}$$

10.5 Radiative transitions in molecules

where T_e is the temperature of the electrons, which are assumed to retain a Maxwellian velocity distribution; T_{bb} is the temperature of the black-body radiation field and W is its geometrical dilution factor. Under the conditions of the interstellar medium, the contribution of stimulated radiative recombination is negligible, either because $W \ll 1$ or because $h\nu \gg k_B T_{bb}$ when $T_{bb} = 2.73$ K, the temperature of the cosmic background radiation. Then, using equation (10.29), we obtain

$$\alpha_\nu(n, T_e) = \frac{\omega_n}{\omega_+} \left(\frac{2}{\pi}\right)^{\frac{1}{2}} \frac{h^3}{c^2} \frac{1}{(m_e k_B T_e)^{\frac{3}{2}}} \exp\left(\frac{h\nu_n}{k_B T_e}\right)$$
$$\times \nu^2 \exp\left(-\frac{h\nu}{k_B T_e}\right) a_\nu(n) \qquad (10.32)$$

The integral rate coefficient for recombination to level n

$$\alpha(n, T_e) = \int_{\nu_n}^{\infty} \alpha_\nu(n, T_e) d\nu \qquad (10.33)$$

can be evaluated when $a_\nu(n)$ is known. The total recombination rate coefficient is obtained by summing over the levels n of the atom A:

$$\alpha(T_e) = \sum_n \alpha(n, T_e) \qquad (10.34)$$

A Fortran program that computes the rate coefficients $\alpha(n, T_e)$ for radiative recombination in the one-electron case is available [207]. The assumption of one active electron is approximately valid also for recombination to *excited* states of many-electron atoms, in which case the 'effective' atomic number is $Z - N$, where Ze is the nuclear charge and N is the number of bound electrons of the recombining ion.

10.5 Radiative transitions in molecules

The probabilities of radiative transitions in molecules can be calculated analogously to those for atoms. However, in addition to the electronic transitions which may accompany the emission or absorption of radiation, account must be taken also of nuclear motions, and hence of changes in the vibrational or rotational state of the molecule. The definition of the *line strength* [equation (10.13) for electric dipole transitions] remains unchanged, but the states i, j involved in the transition are functions not only of the electronic coordinates, $\mathbf{r}_s$, but also of the nuclear coordinate, $\mathbf{r}$, where r is the separation of the nuclei in a diatomic molecule (the case which we shall consider). When the Born–Oppenheimer approximation (see Section 4.2) is used, the electronic and nuclear motions can be separated, i.e. the wave functions of the states i, j can be written as products of electronic and nuclear wave functions. The electronic wave function depends on the electronic coordinates, $\mathbf{r}_s$, in the molecular (body-fixed) frame and also on the magnitude of the internuclear separation, r: the motions of the electrons are assumed to respond instantaneously to variations in r, in such a way as to minimize the total interaction energy for any given value of r. The nuclear wave function is expressed as a product of a vibrational part, dependent on r, and a rotational part, dependent on the orientation of the internuclear axis with respect to the laboratory (space-fixed) frame. Subject

170   *Photon collisions*

to the Born–Oppenheimer approximation, the line strength takes the form

$$S^{v'J'\Lambda'}_{v''J''\Lambda''} = S^{J'\Lambda'}_{J''\Lambda''} p_{v'v''} \tag{10.35}$$

where

$$p_{v'v''} = |\langle v'|M^{\Lambda'}_{\Lambda''}(r)|v''\rangle|^2 \tag{10.36}$$

is the *band strength*, the matrix element of the electronic transition moment, $M^{\Lambda'}_{\Lambda''}(r)$, with respect to the upper and lower vibrational states, v' and v'', respectively; Λ', Λ'' denote the upper and lower electronic states, and J', J'' the upper and lower rotational states. The algebraic quantity $S^{J'\Lambda'}_{J''\Lambda''}$ in equation (10.35) is the *Hönl–London factor*, which determines the selection rules. Formulae for the Hönl–London factors are given by Herzberg [141]; P, Q and R branches are allowed, for electric dipole radiation, corresponding to the selection rule $\Delta J = J' - J'' = -1, 0, 1$, respectively (but $\Delta J = 0$, when $J' = 0$, is not allowed). The Hönl–London factors obey the sum rule

$$\sum_{J''} S^{J'\Lambda'}_{J''\Lambda''} = (2J' + 1) \tag{10.37}$$

The transition probability is given by

$$A(v''J''\Lambda'' \leftarrow v'J'\Lambda') = \frac{64\pi^4 v^3}{3hc^3 \omega_{\Lambda'}(2J' + 1)} S^{v'J'\Lambda'}_{v''J''\Lambda''} \tag{10.38}$$

where v is the transition frequency (v/c is the *wave number* of the transition, the inverse of its wavelength); $\omega_{\Lambda'}$ is the degeneracy (statistical weight) of the upper electronic state, Λ'.

For the (important) case of transitions *within* an electronic state of Σ symmetry, explicit expressions for both the electric dipole and electric quadrupole transition probabilities are given in Section 5.4.1.

10.5.1 Photodissociation of H_2

The photodissociation of H_2 by an ultraviolet radiation field is an important case to which the discussion above relates directly. Photons are absorbed in the Lyman ($B^1\Sigma_u^+ - X^1\Sigma_g^+$) and Werner ($C^1\Pi_u - X^1\Sigma_g^+$) electric dipole transitions; see Fig. 10.1. To the selection rule $\Delta J = 0, \pm 1$ (but $\Delta J = 0$, when $J = 0$, is not allowed) correspond the P, Q and R transitions illustrated in the figure.

The excited electronic states can decay back to the $X^1\Sigma_g^+$ electronic ground state, either to bound rovibrational states or to the vibrational continuum. In the former case, the molecule remains bound, and the subsequent radiative cascade down through the bound rovibrational states can be observed as *fluorescence* radiation. In the latter case, the molecule dissociates. Although this process (of *ultraviolet pumping*, followed by radiative decay to the vibrational continuum of $X^1\Sigma_g^+$), is indirect, proceeding via the $B^1\Sigma_u^+$ or $C^1\Pi_u$ states, it is more efficient than direct photoexcitation to the vibrational continuum. Because H_2 is a homonuclear molecule, it does not possess a permanent electric dipole moment. Consequently, electric dipole transitions *within* the $X^1\Sigma_g^+$ electronic state cannot take place. The lowest non-vanishing permanent multipole moment of H_2 is electric quadrupole, and radiative transitions

10.5 Radiative transitions in molecules

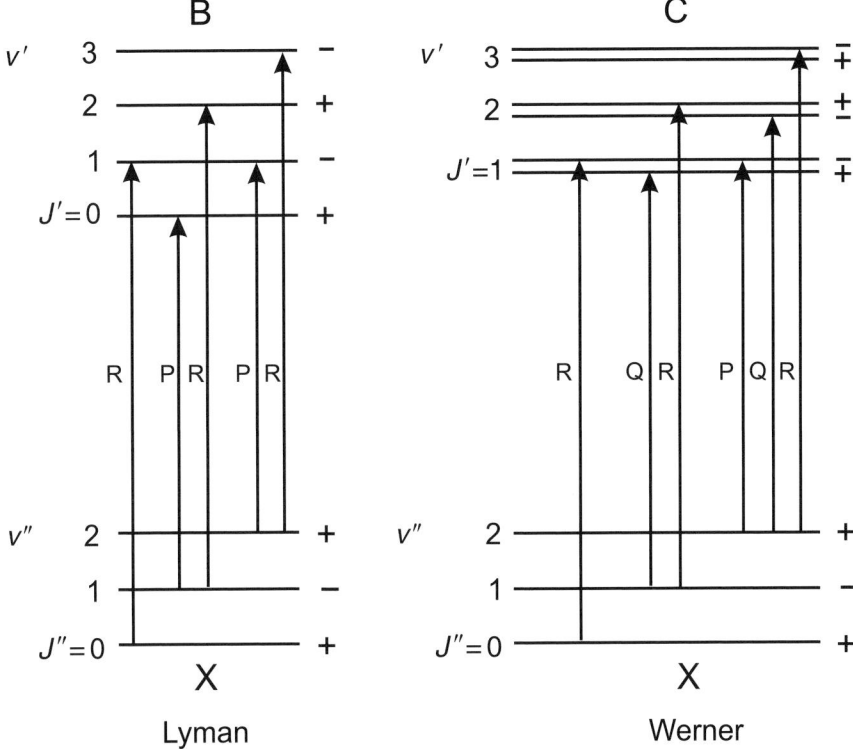

Figure 10.1 Photo-excitation ('optical pumping') of the Lyman and Werner transitions in H$_2$ by an ultraviolet radiation field. The optically allowed transitions from the three lowest rotational states of the vibrational state v'' are illustrated. Transitions in which the rotational quantum number changes by $\Delta J = -1, 0, +1$ are denoted P, Q, R, respectively. The parity of the states is also shown.

within $X^1 \Sigma_g^+$ have small probabilities (and obey the selection rules $\Delta J = 0, \pm 2$; but $\Delta J = 0$, when $J = 0$, is not allowed).

The probabilities of photodissociation of H$_2$ have been computed [208] and the spectrum of H$_2$ in the Lyman and Werner bands has been analyzed [209, 210] by Abgrall *et al.* These calculations were extended subsequently to other excited electronic states [211]. Perturbations to the spectrum occur when rovibrational states belonging to B and C have similar energies. The rovibrational wave functions contain contributions from both the B and the C states, and this 'mixing' of the states becomes significant when the corresponding energy levels are close. The mixing represents departures from the Born–Oppenheimer approximation, which treats the electronic and nuclear (vibrational and rotational) motions as being separable.

Appendix 1 The atomic system of units

In atomic units, the electron charge and mass, e and m_e, and the reduced Planck's constant, $\hbar = h/(2\pi)$, are set equal to unity: $e = m_e = \hbar = 1$. In these units, the radius of the first Bohr orbit in hydrogen

$$a_0 = \frac{\hbar^2}{m_e e^2} = 1$$

becomes the unit of length. The time for $(2\pi)^{-1}$ revolutions in the first Bohr orbit

$$\tau_0 = \frac{\hbar^3}{m_e e^4} = 1$$

is the unit of time. The speed of the electron in the first Bohr orbit is also equal to unity.

The atomic unit of energy is the hartree

$$\frac{e^2}{a_0} = 1$$

which is equal to $2R_\infty$, where R_∞ is the Rydberg constant for infinite nuclear mass (and the reduced mass of the orbital electron is identical to its proper mass). When atomic units are used, the fine structure constant, α, is related to the velocity of light *in vacuo*, c, by

$$c = \frac{e^2}{\hbar \alpha} = \frac{1}{\alpha} \approx 137$$

Appendix 2 Reaction rate coefficients

The rate coefficient for a two-body reaction for which the cross-section is $\sigma(v)$, where v is the magnitude of the relative collision velocity, is given by

$$\langle \sigma v \rangle = \int_0^\infty \int_0^\infty \sigma(v) v f(v_1, T_1) f(v_2, T_2) \mathrm{d}v_1 \mathrm{d}v_2 \qquad (A2.1)$$

in which $f(v_i, T_i)$ denotes the Maxwellian velocity distribution of the reactants, $i = 1$ and $i = 2$,

$$f(v_i, T_i) = 4\pi \left(\frac{m_i}{2\pi k_B T_i} \right)^{\frac{3}{2}} v_i^2 \exp\left(-\frac{m_i v_i^2}{2 k_B T_i} \right) \qquad (A2.2)$$

In (A2.2), m_i is the mass and v_i the velocity of reactant i. Equation (A2.2) may be written alternatively in terms of cartesian velocity components, v_{x_i}, v_{y_i}, and v_{z_i},

$$f(v_i, T_i) \mathrm{d}v_i = \left(\frac{m_i}{2\pi k_B T_i} \right)^{\frac{3}{2}} \exp\left(-\frac{m_i v_i^2}{2 k_B T_i} \right) \mathrm{d}v_{x_i} \mathrm{d}v_{y_i} \mathrm{d}v_{z_i} \qquad (A2.3)$$

where $v_i^2 = v_{x_i}^2 + v_{y_i}^2 + v_{z_i}^2$. Setting $a_i = m_i/T_i$, equation (A2.1) becomes

$$\langle \sigma v \rangle = \frac{(a_1 a_2)^{\frac{3}{2}}}{(2\pi k_B)^3} \int \sigma(v) v \exp\left[-\frac{a_1 v_1^2 + a_2 v_2^2}{2 k_B} \right] \times \mathrm{d}v_{x_1} \mathrm{d}v_{y_1} \mathrm{d}v_{z_1} \mathrm{d}v_{x_2} \mathrm{d}v_{y_2} \mathrm{d}v_{z_2} \qquad (A2.4)$$

where the integral extends over the three cartesian velocity components of each particle.

Now suppose that the velocity distributions are characterized not only by differing temperatures, $T_1 \neq T_2$, but also that the z-components of velocity, v_{z_1} and v_{z_2}, are displaced relative to each other; this situation arises when modelling planar MHD shock

Reaction rate coefficients

waves. The relative velocity of the reactants has components $v_x = v_{x_1} - v_{x_2}$, $v_y = v_{y_1} - v_{y_2}$, and $v_z = v_{z_1} - v_{z_2} + v_d$, where v_d is the relative drift velocity in the z-direction. Equation (A2.4) may now be written

$$\langle \sigma v \rangle = \frac{(a_1 a_2)^{\frac{3}{2}}}{(2\pi k_B)^3} \int \sigma(v) v \exp\left[-\frac{a_1(v_{x_1}^2 + v_{y_1}^2 + (v_z - v'_{z_2})^2)}{2k_B}\right]$$

$$\times \exp\left[-\frac{a_2((v_x - v_{x_1})^2 + (v_y - v_{y_1})^2 + (v_d - v'_{z_2})^2)}{2k_B}\right]$$

$$\times dv_{x_1} dv_{y_1} dv_x dv_y dv_z dv'_{z_2} \tag{A2.5}$$

where $v'_{z_2} = v_d - v_{z_2}$.

The integrals in (A2.5) may be evaluated as follows. Let $a \equiv a_1 + a_2$; then

$$a_1 v_{x_1}^2 + a_2(v_x - v_{x_1})^2 = a\left[v_{x_1} - \frac{a_2 v_x}{a}\right]^2 + \frac{a_1 a_2 v_x^2}{a} \tag{A2.6}$$

and similarly

$$a_1 v_{y_1}^2 + a_2(v_y - v_{y_1})^2 = a\left[v_{y_1} - \frac{a_2 v_y}{a}\right]^2 + \frac{a_1 a_2 v_y^2}{a} \tag{A2.7}$$

Using (A2.6) and (A2.7), the integrals over v_{x_1} and v_{y_1} in (A2.5) may be carried out, yielding

$$\langle \sigma v \rangle = \frac{a^{\frac{1}{2}}}{(2\pi k_B)^2} \left(\frac{a_1 a_2}{a}\right)^{\frac{3}{2}} \int \sigma(v) v$$

$$\times \exp\left[-\frac{a_1 a_2(v_x^2 + v_y^2)}{2a k_B} - \frac{a_1(v_z - v'_{z_2})^2}{2k_B} - \frac{a_2(v_d - v'_{z_2})^2}{2k_B}\right] dv_x dv_y dv_z dv'_{z_2} \tag{A2.8}$$

The integral over v'_{z_2} may be performed by writing

$$a_1(v_z - v'_{z_2})^2 + a_2(v_d - v'_{z_2})^2 = a\left(v'_{z_2} - \frac{a_1 v_z + a_2 v_d}{a}\right)^2 + \frac{a_1 a_2(v_z - v_d)^2}{a} \tag{A2.9}$$

and substituting in (A2.8), which becomes

$$\langle \sigma v \rangle = \left(\frac{a_1 a_2}{2\pi a k_B}\right)^{\frac{3}{2}} \int \sigma(v) v$$

$$\times \exp\left[-\frac{a_1 a_2(v_x^2 + v_y^2 + (v_z - v_d)^2)}{2a k_B}\right] dv_x dv_y dv_z \tag{A2.10}$$

Defining

$$v_{xy}^2 = v_x^2 + v_y^2 = v^2 - v_z^2 \tag{A2.11}$$

Reaction rate coefficients

it follows that

$$dv_x dv_y = 2\pi v_{xy} dv_{xy} \tag{A2.12}$$

and hence that

$$\langle \sigma v \rangle = \left(\frac{a_1 a_2}{2\pi a k_B}\right)^{\frac{3}{2}} 2\pi \int \sigma(v) v \exp\left[-\frac{a_1 a_2 (v_x^2 + v_y^2 + (v_z - v_d)^2)}{2a k_B}\right] v_{xy} dv_{xy} dv_z$$

$$= \left(\frac{a_1 a_2}{2\pi a k_B}\right)^{\frac{3}{2}} 2\pi \int \sigma(v) v \exp\left[-\frac{a_1 a_2 (v^2 - 2v_d v_z + v_d^2)}{2a k_B}\right] v dv dv_z \tag{A2.13}$$

where the integral over v_z extends from $-v$ to v. Performing this integral, the expression for the rate coefficient becomes

$$\langle \sigma v \rangle = \left(\frac{a_1 a_2}{2\pi a k_B}\right)^{\frac{1}{2}} \frac{1}{v_d} \int_0^\infty \sigma(v) v \exp\left[-\frac{a_1 a_2 (v^2 + v_d^2)}{2a k_B}\right]$$

$$\times \left[\exp\left(\frac{a_1 a_2 v_d v}{a k_B}\right) - \exp\left(-\frac{a_1 a_2 v_d v}{a k_B}\right)\right] v dv \tag{A2.14}$$

Defining

$$T_r = \frac{m_1 T_2 + m_2 T_1}{m_1 + m_2} \tag{A2.15}$$

and the reduced mass

$$\mu = \frac{m_1 m_2}{m_1 + m_2} \tag{A2.16}$$

we see that

$$\frac{a_1 a_2}{a} = \frac{\mu}{T_r} \tag{A2.17}$$

and hence obtain

$$\langle \sigma v \rangle = \left(\frac{\mu}{2\pi k_B T_r}\right)^{\frac{1}{2}} \frac{1}{v_d} \int_0^\infty \sigma(v) v$$

$$\times \left[\exp\left(-\frac{\mu(v - v_d)^2}{2k_B T_r}\right) - \exp\left(-\frac{\mu(v + v_d)^2}{2k_B T_r}\right)\right] v dv \tag{A2.18}$$

A useful limit of (A2.18) is attained as $v_d \to 0$, namely

$$\langle \sigma v \rangle \to \left(\frac{\mu}{2\pi k_B T_r}\right)^{\frac{1}{2}} \frac{1}{v_d} \int_0^\infty \sigma(v) v \exp\left(-\frac{\mu v^2}{2k_B T_r}\right) \frac{2\mu}{k_B T_r} v_d v^2 dv$$

$$= 4\pi \left(\frac{\mu}{2\pi k_B T_r}\right)^{\frac{3}{2}} \int_0^\infty \sigma(v) v \exp\left(-\frac{\mu v^2}{2k_B T_r}\right) v^2 dv \tag{A2.19}$$

which is the Maxwellian average of σv, evaluated at the weighted mean kinetic temperature, T_r. Note that, if $T_1 = T_2 = T$, then $T_r = T$.

Consider now an ion–molecule reaction, for which $\mu = m_i m_n/(m_i + m_n)$, $T_r = (m_i T_n + m_n T_i)/(m_i + m_n)$, and $v_d = |u_i - u_n|$, the absolute magnitude of the difference between the flow velocities of the ionized and neutral fluids. If the reaction is endoergic by an amount E_{th}, the corresponding threshold velocity is given by

$$\frac{\mu v_{th}^2}{2} = E_{th} \tag{A2.20}$$

and the range of integration in (A2.18) extends from $v = v_{th}$ to $v = \infty$. Defining

$$s^2 = \frac{\mu v_d^2}{2k_B T_r} \tag{A2.21}$$

$$t^2 = \frac{\mu v_{th}^2}{2k_B T_r} \tag{A2.22}$$

and

$$x^2 = \frac{\mu v^2}{2k_B T_r} \tag{A2.23}$$

equation (A2.18) may be written

$$\langle \sigma v \rangle = \left(\frac{8k_B T_r}{\pi \mu}\right)^{\frac{1}{2}} \frac{1}{s} \int_t^\infty \sigma(x) x^2 \sinh(2sx) \exp(-x^2 - s^2) dx \tag{A2.24}$$

If the energy dependence of the cross-section, σ, is known, the integral in (A2.24) may be evaluated numerically.

References

1. Carruthers, G. R. (1971). Far-ultraviolet spectra and photometry of Perseus stars. *Astrophys. J.* **166**, 349–59.
2. Spitzer, L., Drake, J. F., Jenkins, E. B., Morton, D. C., Rogerson, J. B. and York, D. G. (1973). Spectrophotometric results from the *Copernicus* satellite IV. Molecular hydrogen in interstellar space. *Astrophys. J.* **181**, L116–21.
3. Shull, J. M., Tumlinson, J., Jenkins, E. B., *et al.* (2000). *Far ultraviolet spectroscopic explorer* observations of dissuse interstellar molecular hydrogen. *Astrophys. J.* **538**, L73–6.
4. The *Infrared Space Observatory* (ISO) special issue (1996). *Astron. Astrophys.* **315**(2).
5. Cravens, T. E., Victor, G. A. and Dalgarno, A. (1975). The absorption of energetic electrons by molecular hydrogen gas. *Planet. Space Sci.* **23**, 1059–70.
6. Prasad, S. S. and Tarafadar, S. P. (1983). Ultraviolet radiation field inside dense clouds: its possible existence and chemical implications. *Astrophys. J.* **267**, 603–9.
7. Gredel, R., Lepp, S. and Dalgarno, A. (1987). The C/CO ratio in dense interstellar clouds, *Astrophys. J.* **323**, L137–9.
8. Hollenbach, D. J. and Salpeter, E. E. (1971). Surface recombination of hydrogen molecules. *Astrophys. J.* **163**, 155–64.
9. Hollenbach, D. J., Werner, M. W. and Salpeter, E. E. (1971). Molecular hydrogen in H I regions. *Astrophys. J.* **163**, 165–80.
10. Williams, D. A., Williams, D. E., Clary, D., *et al.* (2000). The energetics and efficiency of H_2 formation on the surface of simulated interstellar grains. In: *Molecular Hydrogen in Space*, ed. F. Combes and G. Pineau des Forêts (Cambridge: Cambridge University Press), pp. 99–106.
11. Morisset, S., Aguillon, F., Sizun, M. and Sidis, V. (2003). Quantum wavepacket investigation of Eley Rideal formation of H_2 on a relaxing graphite surface. *Chem. Phys. Letters* **378**, 615–21.
12. Herbst, E. and Klemperer, W. (1973). The formation and depletion of molecules in dense interstellar clouds *Astrophys. J.* **185**, 505–33.
13. Goldsmith, P. F., Melnick, G. J., Bergin, E. A., *et al.* (2000). O_2 in interstellar molecular clouds. *Astrophys. J.* **539**, L123–27.
14. Bergin, E. A., Melnick, G. J., Stauffer, J. R., *et al.* (2000). Implications of submillimeter wave astronomy satellite observations for interstellar chemistry and star formation. *Astrophys. J.* **539**, L129–32.
15. Graedel, T. E., Langer, W. D. and Frerking, M. A. (1982) The kinetic chemistry of dense interstellar clouds. *Astrophys. J. Suppl.* **48**, 321–68.
16. Larsson, M., Danared, H., Mowat, J.R., *et al.* (1993). Direct high-energy neutral-channel dissociative recombination of cold H_3^+ in an ion storage ring. *Phys. Rev. Lett.* **70**, 430–3.
17. McCall, B. J., Huneycutt, A. J., Saykally, R. J., *et al.* (2003). An enhanced cosmic-ray flux towards ζ Persei inferred from a laboratory study of the H_3^+ recombination rate. *Nature* **422**, 500–2.
18. Pineau des Forêts, G., Roueff, E. and Flower, D. R. (1992). The two chemical phases of dark interstellar clouds. *Mon. Not. Roy. Astron. Soc.* **258**, 45P–47P.
19. Le Bourlot, J., Pineau des Forêts, G., Roueff, E. and Schilke, P. (1993). Bistability in dark cloud chemistry. *Astrophys. J.* **416**, L87–L90.
20. Goldsmith, D. W., Habing, H. J. and Field, G. B. (1969). Thermal properties of interstellar gas heated by cosmic rays. *Astrophys. J.* **158**, 173–83.
21. Black, J. and Dalgarno, A. (1977). Models of interstellar clouds I. The ζ Ophiuchi cloud. *Astrophys. J. Suppl.* **34**, 405–23.

References

22. Hollenbach, D. J. and Shull, J. M. (1977). Vibrationally excited molecular hydrogen in Orion. *Astrophys. J.* **216**, 419–26.
23. Kwan, J. (1977). On the molecular hydrogen emission at the Orion Nebula. *Astrophys. J.* **216**, 713–23.
24. London, R., McCray, R. and Chu, S.-I. (1977). A shock model for infrared line emission from H_2 molecules. *Astrophys. J.* **217**, 442–7.
25. Nadeau, D. and Geballe, T. R. (1979). Velocity profiles of the 2.1 μm H_2 emission line in the Orion molecular cloud. *Astrophys. J.* **230**, L169–73.
26. Field, G. B., Rather, J. D. G., Aannestad, P. A. and Orszag, S. A. (1968). Hydromagnetic shock waves and their infrared emission in H I regions. *Astrophys. J.* **151**, 953–75.
27. Mullan, D. J. (1971). The structure of transverse hydromagnetic shocks in regions of low ionization. *Mon. Not. Roy. Astron. Soc.* **153**, 145–70.
28. Hollenbach, D. J. and McKee, C. F. (1979). Molecule formation and infrared emission in fast interstellar shocks I. Physical processes. *Astrophys. J. Suppl.* **41**, 555–92.
29. Draine, B. T. (1980). Interstellar shock waves with magnetic precursors. *Astrophys. J.* **241**, 1021–38.
30. Smith, M. D. and Mac Low, M.-M. (1997). The formation of C-shocks: structure and signatures. *Astron. Astrophys.* **326**, 801–10.
31. Chièze, J.-P., Pineau des Forêts, G. and Flower, D. R. (1998). Temporal evolution of MHD shocks in the interstellar medium., *Mon. Not. Roy. Astron. Soc.* **295**, 672–82.
32. Lesaffre, P., Chièze, J.-P., Cabrit, S. and Pineau des Forêts, G. (2004a). Temporal evolution of magnetic molecular shocks I. Moving grid simulations. *Astron. Astrophys.* **427**, 147–55.
33. Lesaffre, P., Chièze, J.-P., Cabrit, S. and Pineau des Forêts, G. (2004b). Temporal evolution of magnetic molecular shocks II. Analytics of the steady state and semi-analytical construction of intermediate ages. *Astron. Astrophys.* **427**, 157–67.
34. Draine, B. T. (1986). Multicomponent, reacting MHD flows. *Mon. Not. Roy. Astron. Soc.* **220**, 133–48.
35. Richtmyer, R. D. and Morton, K. W. (1957). *Difference Methods for Initial-value Problems*. (New York: John Wiley & Sons).
36. Osterbrock, D. E. (1961). On ambipolar diffusion in H I regions. *Astrophys. J.* **134**, 270–2.
37. Flower, D. R. (2000). Momentum transfer between ions and neutrals in molecular gas. *Mon. Not. Roy. Astron. Soc.* **313**, L19–21.
38. Draine, B. T. and Sutin, B. (1987). Collisional charging of interstellar grains. *Astrophys. J.* **320**, 803–17.
39. Snell, R. L., Hollenbach, D. J., Howe, J. E., *et al.* (2005). Detection of water in the shocked gas associated with IC 443: constraints on shock models. *Astrophys. J.* **620**, 758–73.
40. Bennett, O. J., Dickinson, A. S., Leininger, T. and Gadéa, F. X. (2003). Radiative association in Li+H revisited: the role of quasi-bound states. *Mon. Not. Roy. Astron. Soc.* **341**, 361–68.
41. Spitzer, L. (1978). *Physical Processes in the Interstellar Medium*. (New York: John Wiley & Sons).
42. Chandrasekhar, S. and Fermi, E. (1953). Problems of gravitational stability in the presence of a magnetic field, *Astrophys. J.* **118**, 116–41.
43. Drake, G. W. F. (ed.) (2006). *Springer Handbook of Atomic, Molecular & Optical Physics*. (New York: Drake (Springer-Verlag)).
44. Arthurs, A. M. and Dalgarno, A. (1960). The theory of scattering by a rigid rotator. *Proc. Roy. Soc. London* **A256**, 540–51.
45. Pack, R. T. (1974). Space-fixed *vs.* body-fixed axes in atom-diatomic molecule scattering. Sudden approximations. *J. Chem. Phys.* **60**, 633–9.
46. Percival, I. C. and Seaton, M. J. (1957). The partial wave theory of electron–hydrogen atom collisions. *Proc. Cambridge Phil. Soc.* **53**, 654–62.
47. Abramowitz, M. and Stegun, I. A. (1965). *Handbook of Mathematical Functions*. (New York: Dover Publications).
48. Rose, M. E. (1957). *Elementary Theory of Angular Momentum*. (New York: John Wiley & Sons).
49. Brink, D. M. and Satchler, G. R. (1968). *Angular Momentum*. (Oxford: Clarendon Press).
50. Edmonds, A. R. (1960). *Angular Momentum in Quantum Mechanics*. (Princeton, NJ: Princeton University Press).
51. Messiah, A. (1969). *Quantum Mechanics, Vol. 2*. (Amsterdam: North Holland Publishing).
52. Launay, J.-M. (1976). Body-fixed formulation of rotational excitation: exact and centrifugal decoupling results for CO–He. *J. Phys. B: At. Mol. Phys.* **9**, 1823–38.
53. Hutson, J. M. and Green, S. (1994). *MOLSCAT version 14*, distributed by Collaborative Computational Project No. 6. (Daresbury Laboratory, UK: Engineering and Physical Sciences Research Council).

References

54. Manolopoulos, D. E. and Alexander, M. H. (1992). Quantum flux redistribution during molecular photodissociation. *J. Chem. Phys.* **97**, 2527–35.
55. Flower, D. R., Bourhis, G. and Launay, J.-M. (2000). MOLCOL: a program for solving atomic and molecular collision problems. *Computer Phys. Commun.* **131**, 187–201.
56. Danby, G. (1983). Theoretical studies of Van der Waals molecules: general formulation *J. Phys. B: At. Mol. Phys.* **16**, 3393–410.
57. McGuire, P. and Kouri, D. J. (1974). Quantum mechanical close coupling approach to molecular collisions. j_z-conserving coupled states approximation. *J. Chem. Phys.* **60**, 2488–99.
58. Tsien, T. P. and Pack, R. T. (1970a). Rotational excitation in molecular collisions: a strong coupling approximation. *Chem. Phys. Letters* **6**, 54–6.
59. Tsien, T. P. and Pack, R. T. (1970b). Rotational excitation in molecular collisions: corrections to a strong coupling approximation. *Chem. Phys. Letters* **6**, 400–2.
60. Tsien, T.P. and Pack, R.T. (1971). Rotational excitation in molecular collisions: a many-state test of the strong coupling approximation, *Chem. Phys. Letters* **8**, 579–81.
61. Pack, R. T. (1972). Relations between some exponential approximations in rotationally inelastic molecular collisions. *Chem. Phys. Letters* **14**, 393–5.
62. Secrest, D. (1975). Theory of angular-momentum decoupling approximations for rotational transitions in scattering. *J. Chem. Phys.* **62**, 710–19.
63. Seaton, M. J. (1962). The theory of excitation and ionization by electron impact. In: *Atomic and Molecular Processes*, ed. D. R. Bates (New York: Academic Press), pp. 374–420.
64. Townes, C. H. and Schawlow, A. L. (1955). *Microwave Spectroscopy*. (New York: McGraw-Hill).
65. Green, S. (1976). Rotational excitation of symmetric top molecules by collisions with atoms: close coupling, coupled states and effective potential calculations for NH_3–He. *J. Chem. Phys.* **64**, 3463–73.
66. Green, S. (1980). Energy transfer in NH_3–He collisions. *J. Chem. Phys.* **73**, 2740–50.
67. Garrison, B. J., Lester, W. A. and Miller, W. H. (1976). Coupled-channels study of the rotational excitation of a rigid asymmetric top by atom impact: (H_2CO, He) at interstellar temperatures. *J. Chem. Phys.* **65**, 2193–200.
68. Green, S., Maluendes, S. and McLean, A. D. (1993). Improved collisional excitation rates for interstellar water. *Astrophys. J. Suppl.* **85**, 181–5.
69. Davis, S. L. and Entley, W. R. (1992). The torsional dependence of an interaction potential: the CH_3OH–He system. *Chem. Phys.* **162**, 285–92.
70. Pei, C. C., Gou, Q. Q. and Zeng, Q. (1988). *Astron. Astrophys. Suppl.* **76**, 35–52.
71. Lees, R. M. (1973). On the E_1–E_2 labeling of energy levels and the anomalous excitation of interstellar methanol. *Astrophys. J.* **184**, 763–71.
72. Lin, C. C. and Swalen, J. D. (1959). Internal rotation and microwave spectroscopy. *Rev. Mod. Phys.* **31**, 841–92.
73. Davis, S. L. (1992). Infinite-order-sudden cross sections for excitation of overall and internal rotation in CH_3OH–He collisions. *J. Chem. Phys.* **97**, 6291–9.
74. Flower, D. R. and Kirkpatrick, D. J. (1982). Rovibrational excitation of molecular hydrogen in collisions with helium atoms. *J. Phys. B: Atomic & Molecular Physics* **15**, 1701–10.
75. Alexander, M. H. and McGuire, P. (1976). Cross sections and rate constants for low–temperature ^{4}He–H_2 vibrational relaxation. *J. Chem. Phys.* **64**, 452–9.
76. Goldflam, R., Green, S. and Kouri, D. J. (1977). Infinite order sudden approximation for rotational energy transfer in gaseous mixtures. *J. Chem. Phys.* **67**, 4149–61.
77. Parker, G. A. and Pack, R. T. (1978). Rotationally and vibrationally inelastic scattering in the rotational IOS approximation. *J. Chem. Phys.* **68**, 1585–601.
78. Muchnick, P. and Russek, A. (1994). The HeH_2 energy surface. *J. Chem. Phys.* **100**, 4336–46.
79. Flower, D. R., Roueff, E. and Zeippen, C. J. (1998). Rovibrational excitation of H_2 molecules by He atoms. *J. Phys. B: At. Mol. Opt. Phys.* **31**, 1105–13.
80. Eastes, W. and Secrest, D. (1972). Calculation of rotational and vibrational transitions for the collision of an atom with a rotating vibrating diatomic oscillator. *J. Chem. Phys.* **56**, 640–9.
81. Dabrowski, I. (1984). The Lyman and Werner bands of H_2. *Can. J. Phys.* **62**, 1639–64.
82. Schaefer, J. and Köhler, W. E. (1985). Quantum calculations of rotational and NMR relaxation, depolarized Rayleigh and rotational Raman line shapes for H_2(HD)–He mixtures. *Physica* **129A**, 469–502.
83. Audibert, M.-M., Vilaseca, R., Lukasik, J. and Ducuing, J. (1976). Experimental study of the vibrational relaxation of ortho- and para-H_2 in collisions with ^{4}He in the range 300–50 K. *Chem. Phys. Letters* **37**, 408–11.

84. Dove, J. E. and Teitelbaum, H. (1974). The vibrational relaxation of H_2 I. Experimental measurements of the rate of relaxation by H_2, He, Ne, Ar and Kr. *Chem. Phys.* **6**, 431–44.
85. Balakrishnan, N., Forrey, R. C. and Dalgarno, A. (1999a). Quantum-mechanical study of ro-vibrational transitions in H_2 induced by He atoms. *Astrophys. J.* **514**, 520–3.
86. Balakrishnan, N., Vieira, M., Babb, J. F., Dalgarno, A., Forrey, R. C. and Lepp, S. (1999b). Rate coefficients for ro-vibrational transitions in H_2 due to collisions with He. *Astrophys. J.* **524**, 1122–30.
87. Flower, D. R. and Roueff, E. (1999). The influence of vibration on rotational cross sections in H_2 and HD. *J. Phys. B: At. Mol. Opt. Phys.* **32**, L171–5.
88. Forrey, R. C., Balakrishnan, N., Dalgarno, A. and Lepp, S. (1997). Quantum mechanical calculations of rotational transitions in H–H_2 collisions. *Astrophys. J.* **489**, 1000–3.
89. Sun, Y. and Dalgarno, A. (1994). Rotational excitation of H_2 in collision with H. *Astrophys. J.* **427**, 1053–6.
90. Mandy, M. E. and Martin, P. G. (1993). Collisional excitation of H_2 molecules by H atoms. *Astrophys. J. Suppl.* **86**, 199–210.
91. Martin, P. G. and Mandy, M. E. (1995). Analytic temperature dependences for a complete set of rate coefficients for collisional excitation and dissociation of H_2 molecules by H atoms. *Astrophys. J.* **455**, L89–L92.
92. Lepp, S., Buch, V. and Dalgarno, A. (1995). Collisional excitation of H_2 molecules by H atoms. *Astrophys. J. Suppl.* **98**, 345–9.
93. Flower, D. R. and Roueff, E. (1998). Vibrational relaxation in H–H_2 collisions. *J. Phys. B: At. Mol. Opt. Phys.* **31**, L955–8.
94. Schaefer, J. (1990). Rotational integral cross sections and rate coefficients of HD scattered by He and H_2. *Astron. Astrophys. Suppl.* **85**, 1101–25.
95. Roueff, E. and Zeippen, C. J. (1999). Rotational excitation of HD molecules by He atoms. *Astron. Astrophys.* **343**, 1005–8.
96. Roueff, E. and Flower, D. R. (1999). A quantum mechanical study of the rotational excitation of HD by H. *Mon. Not. Roy. Astron. Soc.* **305**, 353–6.
97. Flower, D. R. and Roueff, E. (1999). Rovibrational excitation of HD in collisions with atomic and molecular hydrogen. *Mon. Not. Roy. Astron. Soc.* **309**, 833–5.
98. Roueff, E. and Zeippen, C. J. (2000). Rovibrational excitation of HD molecules by He atoms. *Astron. Astrophys. Suppl.* **142**, 475–97.
99. Schaefer, J. and Meyer, W. (1979). Theoretical studies of H_2-H_2 collisions I. Elastic scattering of ground state para- and ortho-H_2 in the rigid rotor approximation. *J. Chem. Phys.* **70**, 344–60.
100. Monchick, L. and Schaefer, J. (1980). Theoretical studies of H_2-H_2 collisions II. Scattering and transport cross sections of hydrogen at low energies: tests of a new ab initio vibrotor potential. *J. Chem. Phys.* **73**, 6153–61.
101. Danby, G., Flower, D. R. and Monteiro, T. S. (1987). Rotationally inelastic collisions between H_2 molecules in interstellar magnetohydrodynamical shocks. *Mon. Not. Roy. Astron. Soc.* **226**, 739–45.
102. Schwenke, D. W. (1988). Calculations of rate constants for the three-body recombination of H_2 in the presence of H_2. *J. Chem. Phys.* **89**, 2076–91.
103. Flower, D. R. (1998). The rotational excitation of H_2 by H_2. *Mon. Not. Roy. Astron. Soc.* **297**, 334–6.
104. Flower, D. R. and Roueff, E. (1998). Rovibrational relaxation in collisions between H_2 molecules I. Transitions induced by ground state para-H_2. *J. Phys. B: At. Mol. Opt. Phys.* **31**, 2935–47.
105. Flower, D. R. and Roueff, E. (1999). Rovibrational relaxation in collisions between H_2 molecules II. Influence of the rotational state of the perturber. *J. Phys. B: At. Mol. Opt. Phys.* **32**, 3399–407.
106. Audibert, M.-M., Joffrin, C. and Ducuing, J. (1974). Vibrational relaxation of H_2 in the range 500–40 K. *Chem. Phys. Letters* **25**, 158–63.
107. Audibert, M.-M., Vilaseca, R., Lukasik, J. and Ducuing, J. (1975). Vibrational relaxation of ortho- and para-H_2 in the range 400–50 K. *Chem. Phys. Lett.* **31**, 232–6.
108. Flower, D. R. and Roueff, E. (1999). Rovibrational excitation of HD in collisions with atomic and molecular hydrogen. *Mon. Not. Roy. Astron. Soc.* **309**, 833–5.
109. Thorson, W. R., Choi, J. H. and Knudson, S. K. (1985). Novel theory of the HD dipole moment II. Computations. *Phys. Rev.* **A31**, 34–42.
110. Poll, J. D. and Wolniewicz, L. (1978). The quadrupole moment of the H_2 molecule. *J. Chem. Phys.* **68**, 3053–8.
111. Le Bourlot, J., Pineau des Forêts, G. and Flower, D. R. (1999). The cooling of astrophysical media by H_2. *Mon. Not. Roy. Astron. Soc.* **305**, 802–10.

References

112. Flower, D. R., Le Bourlot, J., Pineau des Forêts, G. and Roueff, E. (2000). The cooling of astrophysical media by HD. *Mon. Not. Roy. Astron. Soc.* **314**, 753–8.
113. Galli, D. and Palla, F. (1998). The chemistry of the early Universe. *Astron. Astrophys.* **335**, 403–20.
114. Nussbaumer, H. and Rusca, C. (1979). Forbidden transitions in the C I sequence. *Astron. Astrophys.* **72**, 129–33.
115. Nussbaumer, H. and Storey, P. J. (1981). C II two-electron transitions. *Astron. Astrophys.* **96**, 91–5.
116. Wiese, W. L., Smith, M. W. and Glennon, B. M. (1966). *Atomic Transition Probabilities, Vol. I. Hydrogen through Neon*, National Standard Reference Data Series (Washington DC: National Bureau of Standards).
117. Nussbaumer, H. (1977). The Si II spectrum in quasi-stellar objects. *Astron. Astrophys.* **58**, 291–3.
118. Nussbaumer, H. and Storey, P. J. (1980). Atomic data for Fe II. *Astron. Astrophys.* **89**, 308–13.
119. Launay, J.-M. (1977). Molecular collision processes I. Body-fixed theory of collisions between two systems with arbitrary angular momenta. *J. Phys. B: At. Mol. Phys.* **10**, 3665–72.
120. Herzberg, G. (1971). *The Spectra and Structures of Simple Free Radicals* (Ithaca NY: Cornell University Press).
121. Flower, D. R. and Launay, J.-M. (1977). Excitation of the fine-structure transition of C^+ in collisions with molecular hydrogen. *J. Phys. B: At. Mol. Phys.* **10**, L229–33.
122. Buckingham, A. D. (1967). Permanent and induced molecular moments and long-range intermolecular forces. In: *Intermolecular Forces*, ed. J. O. Hirschfelder (New York: Interscience), pp. 107–42.
123. Karl, G., Poll, J. D. and Wolniewicz, L. (1975). Multipole moments of the hydrogen molecule. *Can. J. Phys.* **53**, 1781–90.
124. Kolos, W. and Wolniewicz, L. (1967). Polarizability of the hydrogen molecule, *J. Chem. Phys.* **46**, 1426–32.
125. Weisheit, J. C. and Lane, N. F. (1971). Low-energy elastic and fine-structure excitation scattering of ground state C^+ ions by hydrogen atoms. *Phys. Rev.* **A4**, 171–82.
126. Flower, D. R. and Launay, J.-M. (1977). Molecular collision processes II. Excitation of the fine-structure transition of C^+ in collisions with H_2. *J. Phys. B: At. Mol. Phys.* **10**, 3673–81.
127. Launay, J.-M. and Roueff, E. (1977). Fine-structure excitation of ground-state C^+ ions by hydrogen atoms. *J. Phys. B: At. Mol. Phys.* **10**, 879–88.
128. Launay, J.-M. and Roueff, E. (1977). Fine-structure excitation of carbon and oxygen by atomic hydrogen impact. *Astron. Astrophys.* **56**, 289–92.
129. Wofsy, S., Reid, R. H. G. and Dalgarno, A. (1971). Spin-change scattering of C II and O I by atomic hydrogen. *Astrophys. J.* **168**, 161–7.
130. Green, S., Bagus, P. S., Liu, B., McLean, A. D. and Yoshimine, M. (1972). Calculated potential energy curves for CH^+. *Phys. Rev.* **A5**, 1614–18.
131. Harel, C., Lopez, V., McCarroll, R., Riera, A. and Wahnon, P. (1978). Collision models of intramultiplet transitions at thermal energies. *J. Phys. B: At. Mol. Phys.* **11**, 71–84.
132. Roueff, E. (1990). Excitation of forbidden Si II fine structure transition by atomic hydrogen. *Astron. Astrophys.* **234**, 567–8.
133. Dalgarno, A. and McCray, R. A. (1972). Heating and ionization of H I regions. *Ann. Rev. Astron. Astrophys.* **10**, 375–426.
134. Schröder, K., Staemmler, V., Smith, M. D., Flower, D. R. and Jaquet, R. (1991). Excitation of the fine-structure transitions of C in collisions with ortho- and para-H_2. *J. Phys. B: At. Mol. Opt. Phys.* **24**, 2487–2502.
135. Staemmler, V. & Flower, D. R. (1991). Excitation of the $C(2p^2\ ^3P_J)$ fine structure states in collisions with $He(1s^2\ ^1S_0)$. *J. Phys. B: At. Mol. Opt. Phys.* **24**, 2343–51.
136. Lavendy, H., Robbe, J. M. and Roueff, E. (1991). Interatomic potentials of the C–He system – Application to fine structure excitation of C $^3P(J)$ in collisions with He. *Astron. Astrophys.* **241**, 317–20.
137. Jaquet, R., Staemmler, V., Smith, M. D. and Flower, D. R. (1992). Excitation of the fine-structure transitions of $O(^3P_J)$ in collisions with ortho- and para-H_2, *J. Phys. B: At. Mol. Opt. Phys.* **25**, 285–97.
138. Monteiro, T. S. and Flower, D. R. (1987). Excitation of O I and C I forbidden-line fine structure transitions by He and H_2 – A neglected selection rule. *Mon. Not. Roy. Astron. Soc.* **228**, 101–7.
139. Walmsley, C. M., Batrla, W., Matthews, H. E. and Menten, K. M. (1988). Anti-inversion of the 12.1 GHz methanol line towards dark clouds. *Astron. Astrophys.* **197**, 271–3.
140. Weinreb, S., Barrett, A. H., Meeks, M. L. and Henry, J. C. (1963). Radio observations of OH in the interstellar medium. *Nature* **200**, 829–31.
141. Herzberg, G. (1950). *Molecular Spectra and Molecular Structure I. Spectra of Diatomic Molecules* (Princeton NJ: Van Nostrand).
142. Alexander, M. H. (1985). Quantum treatment of rotationally inelastic collisions involving molecules in Π electronic states: new derivation of the coupling potential. *Chem. Phys.* **92**, 337–44.

References

143. Dewangan, D. P., Flower, D. R. and Alexander, M. H. (1987). Rotational excitation of OH by para-H_2: rate coefficients calculated in an intermediate coupling representation. *Mon. Not. Roy. Astron. Soc.* **226**, 505–12.
144. Larsson, M. (1981). Phase conventions for rotating diatomic molecules. *Physica Scripta* **23**, 835–6.
145. Alexander, M. H. and Dagdigian, P. J. (1984). Clarification of the electronic asymmetry in Π-state Λ-doublets with some implications for molecular collisions. *J. Chem. Phys.* **80**, 4325–32.
146. Dewangan, D. P. and Flower, D. R. (1985). Rotational excitation of OH by H_2: a clarification. *J. Phys. B: At. Mol. Phys.* **18**, L137–40.
147. Dixon, R. N., Field, D. and Zare, R. N. (1985). Collisions of OH and other orbitally degenerate molecules: a consistent treatment of the azimuthal dependence of the interaction potential. *Chem. Phys. Letters* **122**, 310–14.
148. Poynter, R. L. and Beaudet, R. A. (1968). Predictions of several OH Λ-doubling transitions suitable for radio astronomy. *Phys. Rev. Lett.* **21**, 305–8.
149. Brown, J. M., Kerr, C. M. L., Wayne, F. D., Evenson, K. M. and Radford, H. E. (1981). The far-infrared laser magnetic resonance spectrum of the OH radical. *J. Molec. Spectrosc.* **86**, 544–54.
150. Coxon, J. A. and Foster, S. C. (1982). Radial dependence of spin-orbit and Λ-doubling parameters in the $X^2\Pi$ ground state of hydroxyl. *J. Molec. Spectrosc.* **91**, 243–54.
151. Weaver, H., Williams, D. R. W., Dieter, N. H. and Lum, W. T. (1965). Observations of a strong unidentified microwave line and of emission from the OH molecule. *Nature* **208**, 29–31.
152. Litvak, M. M. (1972). Non-equilibrium processes in interstellar molecules, In: *Atoms and Molecules in Astrophysics*, eds. J. R. Carson and M. J. Roberts (New York: Academic Press), pp. 201–76.
153. Gwinn, W. D., Turner, B. E., Goss, W. M. and Blackman, G. L. (1973). Excitation of interstellar OH by the collisional dissociation of water. *Astrophys. J.* **179**, 789–813.
154. Bertojo, M., Cheung, A. C. and Townes, C. H. (1976). Collisional excitation of Λ–doublet transitions in CH and OH. *Astrophys. J.* **208**, 914–22.
155. Dixon, R. N. and Field, D. (1979). Rotationally inelastic collisions of orbitally degenerate molecules: maser action in OH and CH. *Proc. Roy. Soc. London* **A368**, 99–123.
156. Dixon, R. N. and Field, D. (1979). *Lambda*-doublet population inversion in collisions of OH, OD, CH, CD and NH^+. *Mon. Not. Roy. Astron. Soc.* **189**, 583–91.
157. Kaplan, H. and Shapiro, M. (1979). The role of H + OH collisions in pumping the OH 18 cm maser lines. *Astrophys. J.* **229**, L91–6.
158. Shapiro, M. and Kaplan, H. (1979). On the theory of H + OH($^2\Pi$) collisions and interstellar OH maser action. *J. Chem. Phys.* **71**, 2182–93.
159. Dewangan, D. P. and Flower, D. R. (1981). Collisional excitation of OH by H_2: transitions within the ground state Λ-doublet. *J. Phys. B: Atomic & Molecular Physics* **14**, L425–9.
160. Dewangan, D. P. and Flower, D. R. (1981). Rotational excitation of OH by H_2 at thermal energies. *J. Phys. B: Atomic & Molecular Physics* **14**, 2179–90.
161. Dewangan, D. P. and Flower, D. R. (1982). Collisional excitation of OH by H_2 in the interstellar medium. *Mon. Not. Roy. Astron. Soc.* **199**, 457–63.
162. Dewangan, D. P. and Flower, D. R. (1983). Rotational excitation of OH by H_2: calculations in intermediate coupling. *J. Phys. B: Atomic & Molecular Physics* **16**, 2157–68.
163. Rogers, A. E. E. and Barrett, A. H. (1968). Excitation temperature of the 18 cm line of OH in H I regions. *Astrophys. J.* **151**, 163–75.
164. Chu, S.-I. (1976). Collisionally induced hyperfine structure transitions of OH. *Astrophys. J.* **206**, 640–51.
165. Elitzur, M. (1977). Collisional excitations by charged particles in interstellar clouds. *Astron. Astrophys.* **57**, 179–84.
166. Bouloy, D. and Omont, A. (1977). Transitions in Λ-doublets of molecules induced by collions with ions. *Astron. Astrophys.* **61**, 405–10.
167. Bouloy, D. and Omont, A. (1979). Transitions in Λ-doublets of molecules induced by collions with ions – II. *Astron. Astrophys. Suppl.* **38**, 101–18.
168. Johnson, I. D. (1967). A mechanism for maser action of OH molecules in interstellar space. *Astrophys. J.* **150**, 33–45.
169. Elitzur, M. (1979). OH main line masers. *Astron. Astrophys.* **73**, 322–8.
170. Hartquist, T. W. (1979). A comment on a mechanism for pumping OH masers. *Astrophys. J.* **77**, 361–2.
171. Litvak, M. M. (1969). Infrared pumping of interstellar OH. *Astrophys. J.* **156**, 471–92.
172. Lucas, R. (1980). The pumping of interstellar OH main line masers: an efficient mechanism. *Astron. Astrophys.* **84**, 36–9.

173. Guilloteau, S., Lucas, R. and Omont, A. (1981). Puming of H II/OH masers by infrared line overlaps. *Astron. Astrophys.* **97**, 347–58.
174. Wright, M. M., Gray, M. D. and Diamond, P. J. (2004). The OH ground-state masers in W3(OH) I. Results for 1665 MHz. *Mon. Not. Roy. Astron. Soc.* **350**, 1253–71; (2004b). *MNRAS*, **350**, 1272–87.
175. Wright, M. M., Gray, M. D. and Diamond, P. J. (2004). The OH ground-state masers in W3(OH) II. Polarization and multifrequency results. *Mon. Not. Roy. Astron. Soc.* **350**, 1272–87.
176. Gray, M. D. (2003). A comparison of models of polarized maser emission. *Mon. Not. Roy. Astron. Soc.* **343**, L33–5.
177. Kochanski, E. and Flower, D. R. (1981). Ab initio calculations of the OH–H_2 potential energy surface. *Chem. Phys.* **57**, 217–25.
178. Offer, A. R. (1990). Quantal calculations on the rotational excitation of NH_3 and OH in collisions with H_2. Ph.D. thesis, University of Durham.
179. Andresen, P., Haüsler, D. and Lülf, H. W. (1984). Selective Λ-doublet population in inelastic collisions with H_2: a possible pump mechanism for the $^2\Pi_{\frac{1}{2}}$ astronomical OH maser. *J. Chem. Phys.* **81**, 571–2.
180. Schinke, R. and Andresen, P. (1984). Inelastic collisions of OH($^2\Pi$) with H_2: comparison between theory and experiment including rotational, fine structure, and Λ-doublet transitions. *J. Chem. Phys.* **81**, 5644–8.
181. Dewangan, D. P., Flower, D. R. and Danby, G. (1986). Rotational excitation of OH by H_2: a comparison between theory and experiment. *J. Phys. B: At. Mol. Phys.* **19**, L747–53.
182. Offer, A. R. and van Dishoeck, E. F. (1992). Rotational excitation of interstellar OH by para- and ortho-H_2. *Mon. Not. Roy. Astron. Soc.* **257**, 377–90.
183. Condon, E. U. and Shortley, G. H. (1964). *The Theory of Atomic Spectra* (Cambridge: Cambridge University Press).
184. Bates, D. R. (1960). Collisions involving the crossing of potential energy curves. *Proc. Roy. Soc. London* **A257**, 22–31.
185. McCarroll, R. and Valiron, P. (1976). charge transfer of Si^{2+} ions with atomic hydrogen in the interstellar medium. *Astron. Astrophys.* **53**, 83–8.
186. Bates, D. R. and McCarroll, R. (1958). Electron capture in slow collisions. *Proc. Roy. Soc. London* **A245**, 175–83.
187. Gaussorgues, C., Le Sech, C., Masnou-Seeuws, F., McCarroll, R. and Riera, A. (1975). Common trajectory methods for the calculation of differential cross sections for inelastic transitions in atom (ion)-atom collisions I. General theory. *J. Phys. B: At. Mol. Phys.* **8**, 239–52.
188. Baer, M. (1975). Adiabatic and diabatic representations for atom-molecule collisions: treatment of the collinear arrangement. *Chem. Phys. Letters* **35**, 112–18.
189. Heil, T. G. and Dalgarno, A. (1979). Diabatic molecular states. *J. Phys. B: At. Mol. Phys.* **12**, L557–60.
190. Butler, S. E., Heil, T. G. and Dalgno, A. (1980). Charge transfer of multiply-charged ions with hydrogen and helium: quantal calculations. *Astrophys. J.* **241**, 442–7.
191. Watson, W. D. and Christensen, R. B. (1979). Quantal calculations for charge transfer in collisions of C^{+3} and N^{+3} with H atoms. *Astrophys. J.* **231**, 627–31.
192. Gargaud, M., Hanssen, J., McCarroll, R. and Valiron, P. (1981). Charge exchange with multiply charged ions at low energies: application to the N^{+3}/H and C^{+4}/H systems. *J. Phys. B: At. Mol. Phys.* **14**, 2259–76.
193. Butler, S. E. and Dalgarno, A. (1979). Charge transfer between N^+ and H. *Astrophys. J.* **234**, 765–7.
194. Chambaud, G., Launay, J.-M., Levy, B., Millié, P., Roueff, E. and Tran Minh, F. (1980). Charge exchange and fine structure excitation in O–H^+ collisions. *J. Phys. B: At. Mol. Phys.* **13**, 4205–16.
195. Gargaud, M., McCarroll, R. and Valiron, P. (1982). Charge transfer ionization of Si^+ by H^+ at thermal energies. *Astron. Astrophys.* **106**, 197–200.
196. Heitler, W. (1954). *The Quantum Theory of Radiation* (Oxford: Clarendon Press).
197. Baliunas, S. L. and Butler, S. E. (1980). Silicon lines as spectral diagnostics: the effect of charge transfer. *Astrophys. J.* **235**, L45–8.
198. Allan, R. J., Clegg, R. E. S., Dickinson, A. S. and Flower, D. R.(1988). Mg–H^+ charge transfer and Mg line intensities in gaseous nebulae. *Mon. Not. Roy. Astron. Soc.* **235**, 1245–55.
199. Federman, S. R. and Shipsey, E. J. (1983). The ^1D–^3P transition in atomic oxygen induced by collisions with atomic hydrogen. *Astrophys. J.* **269**, 791–5.
200. Shields, G. A., Dalgarno, A. and Sternberg, A. (1983). Line emission from charge transfer with atomic hydrogen at thermal energies. *Phys. Rev.* **A28**, 2137–40.
201. Clegg, R. E. S. and Walsh, J. R. (1985). Charge transfer of O^{3+} with H in planetary nebulae. *Mon. Not. Roy. Astron. Soc.* **215**, 323–33.

202. Clegg, R. E. S., Harrington, J. P. and Storey, P. J. (1986). Ne III charge-exchange lines in the planetary nebula NGC 3918. *Mon. Not. Roy. Astron. Soc.* **221**, 61P–67P.
203. Dalgarno, A. and Sternberg, A. (1982). The excitation of the triplet lines of O^{2+} in nebulae. *Mon. Not. Roy. Astron. Soc.* **200**, 77P–80P.
204. The Opacity Project Team (1995). *The Opacity Project Vol. 1* (Bristol: Institute of Physics Publications).
205. http://vizier.u-strasbg.fr/OP.html
206. Ramsbottom, C. A., Scott, M. P., Bell, K. L., *et al.* (2002). Electron impact excitation of the iron peak element Fe II. *J. Phys. B: At. Mol. Opt. Phys.* **35**, 3451–77.
207. Flower, D. R. and Seaton, M. J. (1969). A program to calculate radiative recombination coefficients of hydrogenic ions, *Computer Phys. Commun.* **1**, 31–4.
208. Abgrall, H., Le Bourlot, J., Pineau des Forêts, G., Roueff, E., Flower, D. R. and Heck, L. (1992). Photodissociation of H_2 and the H/H_2 transition in interstellar clouds. *Astron. Astrophys.* **253**, 525–36.
209. Abgrall, H., Roueff, E., Launay, F., Roncin, J. Y. and Subtil, J. L. (1993). Table of the Lyman band system of molecular hydrogen. *Astron. Astrophys. Suppl.* **101**, 273–321.
210. Abgrall, H., Roueff, E., Launay, F., Roncin, J. Y. and Subtil, J. L. (1993). Table of the Werner band system of molecular hydrogen. *Astron. Astrophys. Suppl.* **101**, 323–62.
211. Abgrall, H., Roueff, E. and Drira, I. (2000). Total transition probability and spontaneous radiative dissociation of B, C, B′ and D states of molecular hydrogen. *Astron. Astrophys. Suppl.* **141**, 297–300.

Index

adiabatic, 16, 21, 23, 30, 36, 37, 44, 46, 47, 51, 98, 106, 107, 111–113, 139–142, 146–149, 183
ambipolar diffusion, 5, 13, 34, 128, 178
ammonia, 6, 69–71, 74, 78
asymmetric top, 69, 74, 76–78, 80, 179
avoided crossing, 141–143, 147, 148

band strength, 170
barycentric, 51, 149
Bessel, 67, 105
bistability, 9, 10, 177
black-body, 36, 44, 122–124, 169
body-fixed, 52, 70, 72, 79, 83, 100, 125, 126, 132–134, 146, 169, 178, 181
Boltzmann, 36, 45, 69, 88, 89, 97, 116, 118, 119, 121, 144
Born–Oppenheimer approximation, 39, 49, 51, 106, 145, 146, 169, 170
boson, 41, 91
boundary conditions, 66, 67, 74, 83, 104
bremsstrahlung, 166

$C_{\infty v}$, 106, 108, 111
C_{2v}, 106, 108, 111
C_s, 107
centrifugal, 52, 59, 60, 63–66, 84–86, 96, 109, 144, 158, 178
CH^+, 2, 9, 17, 34, 124, 181
CH_3OH, 78, 82, 179
charge transfer, 7, 8, 51, 139–152, 183
chemical hysteresis, 9
Clebsch–Gordan, 54–57, 62, 72, 88, 100
CO, 1, 6, 62, 82
collision strength, 69, 160, 161
collisional dissociation, 21–23, 25, 29, 32, 128, 182
cooling, 1, 21–26, 28–30, 42, 45–47, 90, 92–94, 96–98, 107, 115, 116, 180
Coriolis, 52, 60, 64
cosmic background, 36, 45, 78, 118, 169
cosmic ray, 3, 5, 8, 11, 20, 31, 177
coulomb, 5, 33, 89, 91, 140, 141, 153, 156
coupled channels, 66, 67, 85, 86, 89
coupled states, 65, 86, 179
critical density, 7, 39, 45, 46
cross-section, 19, 20, 60, 67, 68, 74, 86–92, 98, 105, 107–110, 113–115, 137, 143–145, 149, 150, 156, 159, 161, 167, 168, 179, 180, 183
cross-section, 68, 69

dark cloud, 3, 8, 9, 26, 78, 139, 177, 181
degeneracy, 69–71, 78, 119, 168, 170
detailed balance, 43, 69, 89, 91, 116, 151, 161
deuterium, 7, 36, 47, 94
diabatic, 98, 141, 142, 148, 149, 183
dielectronic recombination, 149, 151, 160
diffuse cloud, 8, 17, 33, 117, 139
dilution factor, 123, 169
dipole moment, 3, 7, 42, 94–96, 137, 170, 180
Dirac, 70, 77, 110
dissociative recombination, 5–7, 9, 11, 14, 17, 20, 34, 177
Doppler, 37, 120–122, 129, 166

effective potential, 63, 64, 85, 104, 109, 112, 158, 179
Einstein, 38, 39, 119, 120, 164
elastic scattering, 18, 19, 21, 23, 62, 113, 160, 180
electric dipole, 3, 44, 45, 95, 98, 129, 154–156, 165, 166, 169, 170
electric quadrupole, 44, 45, 95, 98, 155, 166, 170
electron, 3, 5, 8, 9, 11, 13, 15, 19, 20, 27, 30, 31, 33, 34, 37–39, 49, 51, 52, 69, 89, 98, 105–107, 110–112, 125, 130–133, 139, 145, 149, 151, 153–169, 177–179, 181, 183, 184
energy conservation, 15, 16, 23, 143
enthalpy, 6, 43
entropy, 24, 43
escape probability, 123
Euler, 53, 71, 72, 125, 126, 134, 146
excitation temperature, 124, 182
exclusion principle, 41, 42, 110, 154, 157

Fermi, 70, 77, 110
fermion, 41, 70, 77, 91, 157
fine structure, 13, 21, 25, 29, 30, 51, 98, 99, 105, 107–110, 112–117, 131, 139, 140, 143, 146, 149, 155, 162, 181, 183
fluorescence, 3, 152
forbidden line, 161, 162
formaldehyde, 74, 76, 77
fractionation, 7, 42, 44, 93, 94, 97

Gaunt, 167
Gaussian, 120
grains, 2–4, 6, 7, 14, 15, 17, 19, 21, 22, 25, 26, 31, 33, 42, 138, 177, 178
gravitational collapse, 45, 46

185

Index

H_2, 1, 3, 4, 6, 8, 12, 13, 15, 17, 19, 21, 23–25, 27, 29–31, 33, 39, 41–45, 47, 48, 86, 89–94, 96–98, 105, 106, 108, 109, 111, 115, 117, 153, 155, 161, 170, 171, 177–182, 184
H_2CO, 179
H_2O, 6, 11, 30, 34, 35, 69, 93
Hönl–London, 170
harmonic oscillator, 40, 81, 84, 89, 164
HD, 7, 39, 40, 42–45, 47, 48, 86, 91–97, 180
heating, 12, 13, 20, 21, 25, 44–48, 128, 181
Heitler, 3
Herbig–Haro, 151, 161, 162
Hubble, 37
Hund, 125, 126
hydroxyl, 5, 74, 80, 128, 182
hyperfine, 128, 129, 182
H II region, 12, 124, 138, 149

impact parameter, 65, 68, 109, 143, 144, 158, 159
inelastic scattering, 89, 160, 179
intermediate coupling, 125, 126, 137, 181, 182
interstellar chemistry, 9, 177
interstellar medium, 1–4, 11, 12, 21, 36, 42, 92, 107, 118, 128, 145, 151, 166, 169, 178, 181–183
interstellar molecules, 7, 8, 69, 74, 78, 124, 182
inversion motion, 71, 82
inversion operator, 56, 57, 101
ionization, 3, 5, 6, 8–14, 16, 19, 20, 26, 27, 30, 31, 33, 139–141, 149–151, 153, 159, 161, 162, 166–168, 178, 179, 181, 183
IOS, 66, 87, 88, 179

Lambda-doublet, 125, 129, 137, 138, 182, 183
Landau–Zener, 140–143, 151
Langevin, 19, 43, 144
Legendre, 53, 58, 60
line strength, 165, 169, 170
Lorentz, 164, 166
LS-coupling, 98, 154, 155, 158, 161, 162
LVG, 121
Lyman, 1–3, 20, 166, 170, 171, 179, 184

Mach number, 23, 24, 28, 29
magnetic dipole, 98, 155, 166
magnetic field, 12, 13, 15–17, 21, 22, 24–26, 28–32, 34, 35, 129, 178
magnetic precursor, 17, 21, 29, 31, 178
magnetosonic speed, 16, 17, 31, 32
maser, 78, 118, 119, 123, 124, 128–130, 138, 182, 183
mass conservation, 14
Maxwell, 68, 88, 116, 149, 168, 169
methanol, 69, 74, 78–81, 118, 119, 179, 181
MHD, 13, 15, 17, 21, 27, 29, 34, 178
molecular cloud, 1, 3–5, 8, 11, 12, 78, 81, 82, 92, 118, 177, 178
moment of inertia, 41, 48, 58, 70, 81
momentum conservation, 15
momentum transfer, 15, 18, 19, 178

NH_3, 6, 78, 82, 179, 183

OH, 6, 34, 69, 79, 82, 119, 124–135, 137, 138, 181–183
opacity, 121, 124, 164

optical depth, 118, 121–123
orbiting, 109, 113, 114, 144, 145, 159
oscillator strength, 164, 165

PAH, 31
parity, 56, 57, 61, 86, 93, 98, 100, 101, 105, 125, 126, 129, 137, 154, 155, 158, 160
Pauli, 41, 110, 154, 155, 157
Percival–Seaton coefficients, 62, 93
photoionization, 17, 20, 34, 167, 168
photon, 1–4, 8, 20, 38, 45, 118, 120–123, 128, 138, 163, 166–168, 170
Planck, 69, 120–122, 124, 167
planetary nebula, 82, 139, 149–152, 154, 166, 183
polarizability, 19, 107, 109, 114, 158, 181
population inversion, 119, 128, 129, 138, 182
potential energy curve, 3, 4, 40, 49, 51, 98, 99, 106–108, 111–113, 131, 139–143, 146, 147, 149, 181, 183
primordial gas, 36, 39, 42–45, 48, 97
proton exchange, 41, 44, 77, 78, 81
proton transfer, 6
pumping, 82, 128, 129, 170, 182

QCT method, 91
quadrupole moment, 95, 96, 107, 108, 110, 180

Racah, 61, 103
radiation field, 1–3, 8, 12, 33, 36–38, 44, 45, 47, 48, 78, 118, 122–124, 139, 149, 155, 163, 166, 169–171, 177
radiative recombination, 34, 139, 152, 166–169, 184
radiative transfer, 118, 119, 121, 122, 124, 129
Rankine–Hugoniot, 16, 23
rate coefficient, 19, 21, 42, 43, 45, 68, 69, 88–94, 114–116, 137, 144, 149–152, 156, 159, 161, 167, 169, 173, 180, 181
reactance matrix, 74, 105
reduced mass, 19, 34, 40, 42, 50, 52, 58, 69, 83, 94, 101, 109, 143, 145, 158
relativistic, 162, 163
resonance, 150, 151, 158–161, 182
rigid rotor, 52, 58, 65, 70, 71, 180
rotation matrix, 53, 54, 63, 71, 76, 101, 125, 134, 135, 146
rotational constant, 41, 42, 59, 66, 74, 77, 78, 84, 85, 89, 94, 96, 126
rotational excitation, 45, 51, 62, 66, 69, 71, 82, 89, 91, 92, 99, 129, 137, 149, 178–183

Saha, 167, 168
scattering matrix, 74, 87, 105, 149
selection rule, 3, 129, 154, 155, 170, 171, 181
shock wave, 4, 11–17, 21, 24–35, 48, 78, 92, 93, 153, 178
source function, 49, 70, 83, 99, 100, 121–123, 125, 126, 136, 169, 178
space-fixed, 50, 52, 146
spherical harmonic, 41, 53, 54, 56, 60, 63, 72, 93, 100, 106, 132, 134, 135
spin, 3, 4, 41, 70, 77, 78, 92, 98, 103, 110, 112, 125, 127, 130, 131, 137, 140, 154–158, 160, 161, 168
spin–orbit, 98, 126, 155, 182
spontaneous, 94, 116, 119, 120, 123, 138, 165, 184
star formation, 128, 151, 153, 177

Index

stimulated, 119, 120, 123, 164, 168, 169
sum rule, 165, 170
symmetric top, 69–72, 74, 76–80, 179

thermal balance, 4, 20, 36, 44, 93, 98
thermal conduction, 21, 23, 24
thermodynamic equilibrium, 3, 46, 118, 121, 124, 127, 129, 167, 168
torsional motion, 69, 74, 79, 80, 82
transition probability, 68, 94, 95, 99, 113, 123, 165, 166, 170, 184
transmission matrix, 87, 113

vibrational excitation, 4, 15, 45, 65, 66, 82, 86, 89, 92, 179, 180
viscosity, 16, 21, 23–25
Voigt, 166

water, 5, 34, 35, 178, 182
Werner, 1, 3, 170, 171, 179, 184
Wigner, 54, 61, 93, 103, 137

X-ray, 3, 12

Zeeman, 129